HOW TO SAVE THE WORLD

Environmental Edition

Peter Hamata, PMP, EP

ISBN (Paperback):979-8-89122-622-7

ISBN (Hardcover): 979-8-89122-623-4

Table of Contents

Acknowledgement

First off, I will say thank you the O.G. Editor-In-Chief Grandma Karen for helping me clean up the rough drafts and convincing me that this manuscript is worth publishing. I will also thank my parents for pushing me to complete my academic studies and guiding me onto this very interesting and challenging environmental career path as well as thank Muna and the kids for giving me a purpose for writing this book. Finally, I want to thank all those whose research I have used to help put this book together and I hope everyone who reads this book will make note the authors mentioned in the "Sources" section as they have done a lot of amazing work in the environmental field.

About Me

I didn't always want to work in the environmental sector; my original interest was in mining. I was always intrigued by the mining industry and how it could move mountains, create lakes, and divert rivers. I was also very amazed at how we can execute projects like developing massive open pits or extensive underground tunnels specially designed to extract the required minerals for our society to function while minimizing the waste material for minimum impact on the environment and maximum efficiency. While I was in school for mining and mineral exploration at the British Columbia Institute of Technology (BCIT), the need for environmental support in the mining industry became very apparent, and I changed my focus from mineral exploration and mine development to supporting the mining sector from an environmental aspect. I understand that we need mined minerals to function and develop as a society for things like batteries for electric vehicles, cell phones, computers/technology, solar panels, satellites, and steel and aggregate for building materials, etc., but new methods and technologies could be developed and used to decrease the adverse effects of mining activities.

Once I completed the mining and mineral exploration program, I went into the Environmental Engineering Technology (EET) program at BCIT. The EET program was very interesting and covered many aspects, including climate change, site assessments, site remediation, risk management, waste management, air quality, and toxicology. The program required a final research project; my original proposal was for exploring something called Fenton's Reagents to Oxidize Polycyclic Aromatic Hydrocarbons (PAHs) in PAH contaminated soils. Unfortunately, the program head felt it was too hazardous as it was using highly concentrated Hydrogen Peroxide to oxidize organic material, so I ended up going with "Exploring the Effectiveness of Biosolids from Metro Vancouver's Municipal Sewage Treatment Process in the Biodegradation of PAH Contamination in Marine Sediments." Some useful findings were discovered, but overall, it was determined that further research is required. I tried to have it published, but since I did not set up my statistics section and trials properly, I have not had any luck getting any journals to publish it; I think I am up to about five people who have read it so far.

My previous work experience was mainly in hazardous materials (asbestos/lead/mould) abatement projects, heavy industrial construction (pile driving), residential construction, security at nightclubs, and working in a kitchen before I decided to go back to school and improve my academic portfolio. During my time in school, I started my own hazardous materials abatement and consulting company, which I ran for almost six years to help pay for my schooling, with the intention of working in the mining industry with an emphasis on environmental work upon completion of the program. The plan fell through when I graduated, the price of gold had decreased to around $1,400 USD/oz, and was continuing to decrease, which meant mining operations were slowing down, and work was hard to find, which forced me into getting a job in the industrial waste management industry to start my career in the environmental sector.

I enjoyed working with hazardous and non-hazardous industrial waste as it is a very dynamic work environment, which can require sampling, reviewing Material Safety Data (SDS) documents, or lab results to understand the characteristics of the waste sludge, wastewater, contaminated material, or waste product. Once you have a strong idea of what the waste stream is, you then use a bit of chemistry and common sense to figure out how to neutralize, solidify, safely handle, package, store, and transport the waste. The clients we serviced worked in a wide range of sectors, including chemical manufacturing, schools, hospitals, forestry, mining, and municipalities, but the largest customers were in the oil and gas sector. I was very impressed with the amount of effort and money the local refineries put into ensuring they were continuing their operations without causing any adverse effects to the surrounding environment.

This is true for the refineries in Metro Vancouver, British Columbia, Canada, but I cannot speak to other operations elsewhere in the world. I eventually moved from the transfer station into the office as a project coordinator/inside sales representative to help with contaminated site remediation projects. I spent my days looking at soil sample lab results, figuring out if the contaminated material is classified as a residential level, industrial/commercial level, or hazardous level. Once the material was properly classified, I would find the proper remediation facility or disposal site, line up the transportation, and set up a "cradle to grave" process for the client.

Just like the hard rock mining sector, the oil and gas sector can be volatile, and when the price of oil dropped to below $40 USD / barrel, the market drastically slowed down, decreasing the amount of work and required workers, which caused massive layoffs, with me being one of them. The worst part of it was that I had just purchased a condo in East Vancouver and got the keys a week before I got laid off, so I was now in a position where I had a mortgage to pay and no job. Luckily, the company gave me a fair severance, so I was not completely in a bad spot and was able to complete the minor renovations on my small East Vancouver condo to make it more livable and pay the bills while I looked for employment.

Seeing as the entire industrial waste management industry was down, I went back to my other area of expertise: hazardous materials abatement. I reached out to a well-respected consulting firm in the Vancouver area, which was based out of the Vancouver International Airport (YVR). I had originally obtained my hazardous materials abatement training through them. They knew of my work in the hazardous materials abatement industry, and they decided to take me on to help with the abatement activities for the upgrades to YVR, as they were helping with the seismic upgrades and airport expansion projects. I was employed with the environmental consulting firm for almost a year, where I was testing for asbestos and lead for abatement to ensure the upgrades could happen safely, and monitoring the abatement projects through air monitoring and visual inspections to ensure people at YVR were safe (they always were). I was eventually headhunted for a role with a major international firm that had recently acquired a local industrial waste management company and wanted to use my experience and contacts to help expand its clients and portfolio. The offer was one I could not refuse, so I made the move over.

I really enjoyed being back in the industrial waste management industry and the unique challenges that come along with it. At this time, to further help the environmental community by getting the company's name and services out to the professionals who would require our services, I joined the Board of Directors for the British Columbia Environmental Industry Association (BCEIA). With the new connections from volunteering with the BCEIA, my pre-existing network, attending multiple conferences where environmental issues and solutions were discussed, and informal meetings with many industry professionals, I was able to bring in a lot of new business to the company. I was included in many interesting projects regarding industrial waste management.

With all this success, I earned myself a spot in both the "President's Club" and "Captain's Club" for being both a top performer as well as showing the strongest leadership qualities in my area. Eventually, I was again headhunted for a role with a local utilities company to help with their asbestos management issues, and again, they made me an offer I could not refuse, so I made the move. During this time, I also decided to obtain a couple designations to help showcase my work experience and academic accomplishments in both the project management and environmental fields, with the Project Management Professional (PMP) designation from the Project Management Institute as well as the Environmental Professional (EP) designation from the Environmental Careers Organization of Canada, with a focus on Waste Management and Environmental Management. I am also currently working toward my Engineer in Training (EIT)/Professional Engineer (P.Eng.) designations. I currently must take four tests to meet their academic requirements.

Being back in the asbestos game while working for the utilities company was interesting, especially all the traveling to remote spots around British Columbia and having to figure out solutions for each of the individual projects. I spent just over four years completing abatement projects and helping the company meet WorkSafeBC and internal expectations as a contracted Project Manager, but my contract was terminated in January 2024 as "my role was not required anymore." With my newfound free time, I decided to put together this book, as I felt I needed to share my findings and ideas from working in the environmental sector, working with environmental professionals, learning about environmental issues and solutions at conferences, and through my own independent research to help fix the environmental issues we have created, hopefully reverse the impacts we have had on the environment, and prolong the time we have living as we are used to.

About the Book

I have been working on this book for approximately eight years. Most of the time was spent passively collecting research, tweaking the layout in my head, creating and improving on my innovative technologies and methods, and daydreaming of the world that would result from the successful completion of the projects. I have been discussing environmental issues with my family and friends for years while providing potential solutions, and it was always brought up that I should make the solutions public information, but I never really took it seriously until now.

This book is my attempt to summarize, at a high level, the major issues we are currently facing, along with solutions currently being explored by the scientific community, solutions I feel would work, and potential solutions that could hopefully be improved upon by others and eventually get to a point of success. The book is written for people who are interested in learning about the major environmental issues that environmental professionals are currently dealing with, as well as what we are doing and must do to get the environment back to a sustainable point. This book is also meant for the philanthropists out there who have the funds to help reverse the damage we have caused to the environment, who want to do their part, and leave the legacy of being the person who helped save the planet.

I have done the research, so you won't have to. Some of the findings came naturally through working in the environmental industry with environmental professionals, while others came through my own independent research from my genuine interest in finding solutions for the environmental issues we have created. There is a lot of really great work that is currently happening, especially in the energy creation/storage sectors, farming, carbon capture, and sustainable construction. A fantastic resource I can guide you toward, if you are interested in learning more about current solutions, is a book by Tom Heap titled "39 Ways to Save the Planet" (thanks, Mom).

The great thing about reversing the environmental impacts we have on the environment is that it is something that everyone benefits from. The only real pushback from implementing the environmental projects required to reverse the negative effects would be from companies that have their profits reliant on operations that are harmful to the environment. This book will hopefully provide some solutions, or at least start some discussions on how the companies can alter their operations to be more sustainable, cause less impact to the environment, and still make a profit. Just as I am interested in helping the mining sector continue its operations to extract the required minerals for our society to function in an environmentally sustainable way, I am interested in helping other industries continue to meet the demand of our society by producing their goods and services, but doing it in a more sustainable way. I think that there are some industries that need to be reduced or eliminated, just as we did with producing asbestos-containing material, certain lead and mercury-containing materials, as well as using alternate (more sustainable materials) for material production.

Some of the solutions provided in this book are already being researched or implemented, but I feel they need more attention for further research and development, or maybe just more funding. Some of the solutions designed by me are ideas that could be built and improved on, but they are discussed in this book to help showcase a pathway to the solutions for the issues.

As previously mentioned, the environmental projects require funding. The major issue with the environmental cleanup projects is that there is no direct monetary return on investment (ROI) for investors, so many people have no interest in putting money towards it when they can invest in something that will make them more money. There are tax credits and improved reputation, which both can get more money in your pocket, but many financially secure people prefer to keep using their money to make more money, and funding environmental projects does not offer a good monetary return. Imagine a world where people idolized individuals and organizations who gave up a small portion of their accumulated wealth to fund environmental projects instead of idolizing people who just spend their money on expensive cars, boats, jets, homes, clothes, and jewelry just for attention and bragging rights; I bet you there would be enough funding to fix all of the environmental issues in a couple of years if that were the case.

Environmental issues are global issues, and we need global solutions to tackle them. It is great how some countries have signed onto protocols, accords, and agreements to help combat climate change and pollution, but it seems like many of them, including th_e ones that have the largest impacts, are not interested and have backed out. They are not able to meet their goals, or could just change their mind and back out at any time. The risk of governments not meeting their environmental goals is high due to instability with government relations, newly elected leaders changing policies, political/global issues, etc.; this shows the importance of placing the responsibility of fixing the issues onto the private sector. This book will hopefully provide a roadmap on what projects we need to execute, technologies we need to research further, societal changes we need to make to get the environment back to where it needs to be; it will also show the need for environmental champions who are willing to sacrifice some of their wealth to help save the world and leave an epic legacy.

Background

Greenhouse Gases (GHGs)

GHGs are gases that trap a portion of heat from the Sun and human activities close to the surface of the Earth and prevent the heat from escaping back into outer space. GHGs help create a livable environment, and they are why the surface of the Earth is much warmer than other planets. The five top GHGs are as follows:

Carbon Dioxide (CO_2):

Carbon Dioxide is the most abundant GHG (67%). It is naturally occurring from sources such as wildfires and volcanoes, but also from human activities like extracting and burning fossil fuels. The Earth was able to develop as it did and support life mainly due to CO2 forming naturally, and over time, it trapped enough of the Sun's heat, creating livable conditions. Since the Industrial Revolution in the 18th century, human activity has drastically increased the amount of CO2 in our atmosphere. This created a situation that is causing too much of the Sun's heat to be captured close to the surface of the Earth and warming our planet up at a harmful rate. According to NASA, CO2 can stay in the atmosphere for 300-1,000 years..

Methane (CH_4):

Methane is not as abundant as Carbon Dioxide (16%) but is 28 times better at trapping heat close to the surface of the Earth and warming it up. Methane naturally occurs from organic material decomposing in environments that do not have oxygen, and human activities have increased the amount of processes that produce methane. Over the past 200 years, the methane amounts in our atmosphere have doubled, mainly due to human activities, such as disposing of organic waste in landfills, agriculture, fossil fuel extraction, and the accelerated thawing of the permafrost in the polar regions of Earth. According to the International Energy Agency, methane can stay in the atmosphere for up to 12 years.

Nitrous Oxide (N_2O):

Nitrous Oxide is also a naturally occurring gas, but with an ability 265 times stronger than CO2, it is a very harmful GHG (6%). Human activities that have greatly increased the amount of Nitrous Oxide to extremely harmful levels are from fertilizer for farming, fossil fuel combustion, and sewage treatment. According to the U.S Environmental Protection Agency (EPA), nitrous oxide can stay in the atmosphere for 121 years.

Fluorinated Gases (F-gases):

These gases are strictly produced by human activity, as they are used for keeping things cold (2%). There are many different types of fluorinated gases, but the main ones are Chlorofluorocarbons (CFCs) and hydrochlorofluorocarbons (HCFCs), which can be found in household and industrial refrigerators and air-conditioning units, as well as Sulfur Hexafluoride (SF6), which is found in industrial circuit breakers in places like substations. These are also known as ozone-depleting gases. According to the U.S. EPA, depending on the type of Fluorinated Gases, they can last for a few weeks to thousands of years in the atmosphere.

Water Vapor:

Water vapor is, of course, naturally occurring, but is also created in higher amounts from industrial operations and manufacturing. Water vapor's ability to trap heat and prevent it from escaping our atmosphere is notable and should be taken into consideration when discussing GHGs. This GHG does break down quickly, but since many industrial operations generally do not stop producing steam (water vapor), the steam you see leaving the "smokestacks" of industrial facilities does play a role in the accelerated heating up of our planet.

Carbon Sinks

Carbon sinks are natural or human-made environments that remove and store CO2 from the atmosphere. Before the Industrial Revolution and the drastic addition of GHGs to the atmosphere, the planet had a healthy balance between the production of GHGs and their removal, which created a livable environment. Major natural carbon sinks are forests, wetlands, and the oceans. New technologies have been invented to help remove CO2 from our atmosphere, but this technology is very expensive and is not very efficient at this time, and requires more research and development.

Circular Economy

This concept is a system that utilizes the waste material from one process and uses it as a product in another process to reduce the amount of waste generated as well as the number of raw materials required to produce new products.

Photosynthesis

The process by which green plants and certain other organisms transform light energy into chemical energy is called photosynthesis. During photosynthesis in green plants, light energy is captured and used to convert water, carbon dioxide, and minerals into oxygen and energy-rich organic compounds (carbohydrates). The chemical reaction that occurs during photosynthesis is as follows: Carbon Dioxide + Water = Carbohydrate and Oxygen ($6CO2 + 6H2O \rightarrow C6H12O6 + 6O2$). Later in the book, we talk about how trees and vegetation release carbon when they decompose or burn, and it is the carbon from the carbohydrates stored in their leaves, trunks, and roots that is released as it breaks down.

Cloud Seeding

The process of manipulating the atmosphere to help form clouds and produce precipitation, like rain or snow, is called cloud seeding. It is a process that has been used and studied since the 1940s and can be done by introducing silver iodine or materials that attract water, like table salt, into the atmosphere to help attach the water molecules together to form clouds and hopefully create rain or snow. Technological advancements have been made that use drones and electrical charges or infrared laser pulses to help promote cloud growth and precipitation in areas that need rainfall. This process can be harmful to the environment if materials like silver iodide are used, as they can make their way down to the surface of the Earth and contaminate people and wildlife; using lasers and electrical charges is a much safer process to accomplish this.

Major Natural Influences on the Environment

It is quite amazing how our planet developed an environment suitable for life as we know it, and there are many natural factors that contributed to it. The scientific community believes that the first thing that had to happen was that the Earth had to cool down from its molten rock phase. At this time, it was believed that volcanoes were abundant, releasing GHGs like CO2 and water vapor into the air, creating an atmosphere and trapping heat close to the surface of Earth, which enabled it to create a temperature suitable for life. It is also believed by the scientific community that the Earth was bombarded by asteroids, which carried and deposited water onto Earth. Once the Earth had enough liquid water bodies and an atmosphere to create a warm enough environment, bacteria started to grow, and oxygen was created through photosynthesis. Over a couple of hundred million years, the atmosphere developed to the point it is today with the current air composition. Earth has a natural cycle to it where it gets cold and warms up again; it has already gone through five major ice ages, and humans have been around for and survived two of them. Within those major ice ages, smaller ice ages occur (called glacials), when ice caps grow for about 90,000 years and then melt (called interglacials) for about 10,000 years. We are currently at the tail end of an interglacial phase and should be entering a glacial phase, but the Earth's natural cycle has been disrupted by global warming via human activity, and it is struggling to start forming the ice caps again. The Earth was naturally able to maintain a reliable cycle due to many factors including the Earth's tilt, which changes from 24.5 degrees to 22.1 degrees, its orbit around the Sun, which changes from a circle to an oval, as well as the Earth's wobble. Other major natural contributing factors that affect the Earth's temperature, weather, and air quality are GHGs, the Sun and ozone layer, polar ice caps, the tropical rain forest, ocean currents, and wildfires. Since we cannot change how the Earth moves in the solar system, we cannot adjust the Sun and we can only control what happens on the surface or close to the surface of the Earth, This book will only focus on that.

Greenhouse Gases

As previously mentioned, greenhouse gases play a very important role in creating a livable environment; they enable the planet to hold heat close to the surface of the Earth and keep it warm enough for life to flourish. Naturally occurring GHGs are created through natural activities like the biological breakdown of organic materials, wildfires, and volcanoes, but they have been kept in check by other natural carbon sinks like forests, wetlands, and oceans, which have maintained a reliable cycle of the global climate until recently.

Sun / Ozone Layer

The Sun is the main source of heat, which is fundamental for the survival of life on Earth. The heat and light are key components in allowing vegetation to perform photosynthesis, removing CO2 from the atmosphere, and creating oxygen for us to breathe. It also provides heat for our planet to have a livable environment, but the Sun's energy also contains ultraviolet (UV) radiation, which in small amounts is beneficial to life but can also be very harmful in large amounts. UV radiation can aid in tasks like synthesizing vitamin D in humans to support bone health, improve disease prevention in humans, and help break down organic material, creating more nutrients for bacteria. It can also alter DNA in plants and animals, inhibit photosynthesis, and kill good bacteria. During the creation of our atmosphere, the ozone layer was naturally formed by gas (oxygen) high above the surface of the Earth (9 to 18 miles or 15 to 30 km) being chemically changed by the Sun's radiation and turning from oxygen (O2) to ozone (O3). The ozone molecules help filter out some of the Sun's UV radiation and have been letting just enough through to support life, but recently, with the creation of ozone-depleting substances like fluorinated gases and aerosols, the ozone layer has been reducing, letting more UV radiation through and creating a dangerous environment here on Earth.

Ice Caps

The ice caps help control the Earth's temperature by reflecting the Sun's heat back up into space and preventing the Earth from heating up too much. They also help keep the organic material frozen in the ground, preventing it from decomposing and releasing methane. It can do this due to its high albedo (its ability to reflect light) and insulating abilities. With the planet not able to regrow its icecaps, Earth's major natural heat deflector and ground cooling system is failing, which allows the Earth to absorb more of the Sun's heat, increasing the Earth's temperatures, further melting the ice caps and creating a snowball effect, which is very difficult to reverse. This will not only increase the amount of heat the Earth is absorbing, but also increase the amount of methane the Earth releases into the atmosphere, drastically increasing the amount of heat trapped in our atmosphere and further increasing the Earth's temperature. The melting of the Earth's ice caps is a major problem as it creates a negative feedback loop, as it creates a warmer Earth, further melting the ice caps.

Ocean / Currents / Weather

The ocean's currents are mainly controlled by the heating and cooling of the ocean water. The surface water is heated at the equator, travels north where it cools down, becomes denser, sinks down, and travels back to the equator at deep levels; this constant current is called the thermohaline circulation or the global circulation conveyor. Scientist's discovered that with the extreme melting of the ice caps, the fresh water is being introduced into the ocean at high rates, making the water less dense, preventing it from sinking and consequently slowing down the ocean currents; this could eventually stop the ocean currents all together, which would create catastrophic changes to oceanic habitats and the weather. Without currents moving the ocean's water around, regional temperatures and weather would be much more extreme, as the heat at the equator being absorbed by the ocean would increase the ocean's temperature, increase the amount of water being evaporated, and increase the severity of bad storms. The warmth from the equator would also not be transferred to the northern and southern regions of the world, and these impacts would make it very difficult to live.

The ocean not only plays a major role in absorbing heat and distributing it around the world, but as previously mentioned, the oceans and the life in the oceans play a major role in capturing carbon from the atmosphere. The oceans cover about 70% of the Earth's surface, and they can absorb carbon for two main reasons. The first is that carbon dioxide can be dissolved in water, and the second is that organisms in the ocean, like phytoplankton and vegetation, absorb CO_2 and use it in their photosynthesis process. Unfortunately, as CO_2 is dissolved in the ocean, it makes the ocean more acidic and less hospitable for life to exist, like phytoplankton and vegetation, which would decrease the amount of CO_2 being absorbed as well as decrease aquatic life.

Amazon Rainforest

Healthy forests generally absorb more carbon dioxide than they emit, making them reliable carbon sinks. Trees and other vegetation do release CO_2 when they decay or are burnt in wildfires, and the decomposition of organic materials can also create methane. Lucky for us, they naturally absorb more CO_2 for photosynthesis than they release, and that also offsets the release of methane. The Amazon Rainforest is by far the largest forest in the world, approximately 6.7 million km^2, spanning eight countries, containing approximately 400 billion trees, and currently storing approximately 150-200 billion tons of carbon. The Amazon rainforest is currently estimated to absorb 25% of the world's CO_2 emissions, and was absorbing 30% more in the 90s before massive deforestation occurred. The rainforest also plays a major role in controlling the climate in South America by storing and recycling water, creating humid climates and rainfall, which helps to create a lush and healthy environment.

In total, it is estimated that 7.6 billion metric tons of CO_2 are absorbed by trees every year, while approximately 36.6 billion tons of CO_2 were produced in 2022. Unfortunately, there is massive deforestation occurring in the Amazon rainforest, mainly from land clearing for agriculture, mining, and dam generation, as well as logging. With the destruction of the rainforest, the amount of trees removing and storing CO_2 is significantly reduced, and large amounts of that previously stored CO_2 are being released. The water which was being stored and recycled in the forest is also being released, disrupting the weather patterns and creating a drier area, creating a less hospitable environment. There are concerns that due to the destruction of the Amazon rainforest, it is now producing more CO_2 and methane than it is extracting and becoming a GHG contributor instead of a vital carbon sink. If we lose the rainforest carbon sink, we will be in a lot of trouble.

Major Human Influences on the Environment

Greenhouse Gases

We have been able to measure and calculate the changes in greenhouse gas emissions from before the Industrial Revolution (approximately 1760 to 1840) to the present day through calculating the amount of GHGs frozen in the polar ice and from modern air sampling technologies. The ice samples are collected by drilling into the ice, obtaining a core sample, and then the scientists collect small samples from the core in spots where they know the date the ice formed and test it for greenhouse gas concentrations. Scientists have found very clear evidence that the concentrations of GHGs in the atmosphere have gone up approximately 40% since before the Industrial Revolution. This makes sense, as the use of coal, oil, and other fossil fuels drastically increased after the Industrial Revolution to support the new industrial operations and sustain our society as we currently know it. The average temperature of the Earth has also been noticed to be increasing steadily due to the increased amounts of GHGs humans have been creating from the new processes we have used since the Industrial Revolution. In 2021, 6.34 million metric tons of CO_2 equivalent were produced in the U.S. from the following activities: 35% from transportation, 31% from generating electricity, 15% from industrial activities, 11% from residential & commercial, and 8% from non-fossil fuel combustion.

Deforestation

Forestry products are needed for our society to function; consequently, logging is an important and useful industry if done properly. I have been involved in multiple arguments regarding logging, and a main concern is not just the destruction of the habitats for wildlife, but also the disruption of the mycelium network, which transfers messages between the trees and vegetation to help them "communicate" and share resources like water and nutrients to where they are needed. Since trees and vegetation naturally die and the mycelium network still recovers and continues to operate, if logging is done properly and only specific trees are removed, the mycelium network will still be able to function normally. I feel like clear-cutting and burning large sections of forest is bad, not only because it would damage the mycelium network and destroy habitats, but also, as mentioned before, because burning the trees, cutting them down, and having pieces of them decompose will release GHGs into the atmosphere, adding to our GHG problem.

Another major argument I've come across is with cutting down old-growth trees and how they should not be logged, as they have been around for a very long time and provide homes for wildlife. In my opinion, based on research, younger trees absorb more CO_2 than older trees, and older trees store more CO_2 than younger trees. If the older trees are left to die and decompose on their own, they would release the stored CO_2, and most of the work they did with capturing CO2 during their lifespan would be in vain. It is my opinion that the ideal solution would be to selectively remove individual old-growth trees (not enough to disrupt the mycelium network and local habitats) and process them in a way to lock in the CO2, then plant new trees that will absorb more CO2.

Unfortunately, what is happening around the world, especially in the Amazon rainforest, is that massive sections of the forest are being burned and cut down for things like agriculture, mining, and forestry products. Sorry to break it to you, but the Western world's obsession with avocados has had a significant negative impact on the environment. The massive increase in demand for avocados has created a need for a massive supply of them. For the supply to be created, large areas of forests had to be destroyed to create farms to make your avocados. The upside is that the increase in demand has created many jobs for people in these developing countries, which enables them to provide a better life for themselves and their families, but massive amounts of forest have been burned or clear-cut just to free up the land. As mentioned before, those activities released large amounts of GHGs into the atmosphere; there is the transportation of the avocados from these countries to your local grocer, which also generates more GHGs.

Another crop that is in high demand and has caused major deforestation for its production is palm oil. Palm oil is found in many common products like food, detergents, fuel, cosmetics, lipstick, soaps, detergents and some ice cream. Large areas of forests have been and are currently being destroyed across Asia, Africa, and Latin America so space can be created for plantations to meet the global demand for palm oil. It is important to know what is in the products you purchase; if we put a demand on products that contain palm oil, we will increase the amount of palm oil needed to be produced. If we decrease our demand for products that contain palm oil, less land will be destroyed for palm oil plantations. If, as a society, we can decrease the demand on things like avocados and palm oil-containing products, or at least have a certification program like we have with coffee to help show that the products were procured by a sustainable source and not a slash-and-burn plantation, we would be in much better shape.

Population Increase

To put it simply, with how we currently operate, the more people that are on the planet, the more resources we need; the resources include products for building homes and businesses, food, clothing, manufactured goods, means for transporting people, areas to live, etc. If everyone was able to just live off the land, eat/shop/trade/live locally and not create massive cities and massive demands on products which are not local and readily available, we would be fine; unfortunately, the odds of that happening are low and we must figure out ways to support the increasing needs of the increasing population while operating sustainably. Developing circular economies would greatly benefit us by reducing the demand for the required raw materials and energy use from the growing population. We should also look at changing how we buy things and our eating habits; these will be discussed later in this book.

Waste

Waste material is an interesting area; it directly contributes to greenhouse gas emissions and can also destroy local habitats. Waste materials include residential solid waste and industrial/manufacturing waste; wastewater is also an ongoing issue from sewage and runoff. Proper waste management is vital to the health of the environment, but many places around the world do not have proper waste management systems, which include collection, treatment, and processes, and this is causing massive destruction to the planet. It is great how many of the developed countries are stepping away from single-use plastics like plastic bags and straws, but the sad thing is that developing countries generally don't have proper waste management systems in place, and are the cause of the global waste management issues. We need the developing countries, which are the main contributors of the loose waste in the environment (land and water), to make enough of a change to tip the scales in our favor.

During my travels, it was apparent that many countries with the largest populations are creating the most waste, but they do not have the resources to set up proper waste management processes. If we are going to help fix the global waste management issue, we need to provide support to these developing countries to help them get to a point where their waste is properly collected, transported, treated and stored. I have noticed in the developing countries that many people have bigger issues like providing food and basic necessities for their families, and that the global waste management issue is not a high priority for them, but you can't blame them for that.

The ideal solution, which is being used in a few regions around the world, is creating waste-to-energy facilities. These facilities burn the waste at such a high temperature that there is minimal smoke. Any smoke that is created is filtered out. The heat generated is used to create steam, which spins a turbine and generates power for consumers. The residual toxic ash that remains from incinerating the waste then requires proper disposal, but the small amounts of hazardous waste created are far less than the positives created, from removing garbage from our societies and the electricity generated. These facilities are very expensive to build and maintain, but we will go into this further later in this book.

Solutions To The Major Issues

Now that you have a general understanding of how the environment works and the current issues we are facing, we can dive into it. Someone once told me about how a solution can be simple, but not easy, and I think many of the solutions to our environmental issues are just that. The simplest solution to almost all the issues related to global warming and extreme climate change is to reduce or stop the production of GHGs entering the atmosphere and remove the excess GHGs. This will require a lot of changes in the world, and it currently seems like many people lack the interest, support, or time to make that happen, which is why I am discussing solutions for not only reducing GHG production, but also projects to help reduce the impacts of a warming planet. Something to keep in mind is that the environment was functioning normally until the Industrial Revolution, which is why many scientists believe that if we can get the GHG concentrations to a level like before the Industrial Revolution, the planet's natural cycles will be back on track. The industrial revolution was not all bad; it pushed humanity to create some amazing technologies, which improved our quality of life. I believe a major driver for the industrial revolution and the post-revolution advancements is mainly to make life easier for everyone by creating more supply, more options, and creating a world where people can simply have anything they want or do anything they want at any time, and create a world where people can live their best lives.

In the project management world, before we start on a project, we must create something called a 'risk registry', which basically is a list of everything that could go wrong with a project and then figure out solutions to either prevent the issues from happening or solutions if the issues occur. There are four main types of risks, the "known knowns" (things that we know), the "known unknowns" (things that we know exist but aren't certain about), the "unknown knowns" (these are the things we don't know that we know), and the "unknown unknowns" (things we do not know we don't know). Planning for the unknown unknowns is very difficult, and the degradation of the environment was an unknown unknown outcome of the Industrial Revolution, as the understanding and concern for the environment were minimal. I can understand how someone could believe back then that, because the world was so big, there was no way things like smoke and other pollutants could alter the planet. Their intent was to make a better world for everyone by advancing our society and technologies, but the unknown unknowns are now a problem, due to the heavy industrial operations, heavy deforestation, and a society reliant on long-distance transportation for everything from food, clothing, vehicles, technology, etc., and we continue to do it.

The current environmental impact now is a "known known" for us, but we continue to proceed to meet the demands of our society. I think the Industrial Revolution was a great start to the advancement of our society, but now it's our turn to start the environmental revolution to help further develop our society into a sustainable one where future generations can flourish and not have to worry about fresh water, clean air and food and still have the luxuries we are used to. We need to change the way we operate to create a sustainable process, starting with the execution of a couple of projects to reverse the damage we have caused and bring our environment back to where it should be. We have the technology and the plans; we just need the champions to step forward to sponsor the projects and get the environmental revolution into full swing.

We have created an environment where the concentrations of GHGs are too high in the atmosphere, and they are preventing heat from escaping the Earth's atmosphere and getting back into outer space. The trapped heat makes the surface of the planet too hot, which causes a vast array of issues like extreme weather conditions, food and water shortages, and the creation of multiple negative feedback loops, which could get the environment to a point of no return and have it spiral out of control. An example of these negative feedback loops is the melting and thawing of the ice caps and permafrost in the polar regions, which then releases more methane from the decomposing organic materials that were previously frozen in the ground. The addition of fresh water being introduced into the ocean from the melting of the ice caps slows down the ocean's currents, preventing the transportation of heat, causing extreme weather events. Basically, the excess heat is causing issues, which causes the planet to heat up more, which then makes the issues worse, causing our planet to heat up even more, and making the issues even worse. This negative feedback loop will continue until we make some major changes.

Reducing Incoming Solar Radiation

This section of the book will discuss what governments around the world are currently doing to try to reduce the amount of incoming solar radiation, and why I disagree with it. I will also discuss two options in this book which I am more in favor of; one will cost a lot more and take more time, but will be more beneficial long term due to the control it will give us on the effects on our planet, and the second option is quicker to implement and costs a lot less money, but we would have no control over it. This section is the most "far-out" section of the book and might seem extreme, but if you can make it through this section, it will be clear sailing.

Catherine Clifford wrote an article published by CNBC on Oct 13, 2022, titled, *"White House is pushing ahead with research to cool Earth by reflecting back Sunlight."* In it, she mentions some of the following. In 2022, the U.S. government created a 5-year research program as part of the Consolidated Appropriations Act for the following reasons:

"Scientific assessment of solar and other rapid climate interventions in the context of near-term climate risks and hazards. The report shall include: (1) the definition of goals in relevant areas of scientific research; (2) capabilities required to model, analyze, observe, and monitor atmospheric composition; (3) climate impacts and the Earth's radiation budget; and (4) the coordination of Federal research and investments to deliver this assessment to manage near-term climate risk and research in climate intervention."

Current processes and technologies, which are being researched and developed by the government and the private sector to help reduce the amount of solar radiation entering our atmosphere, include spreading reflective particles over the ocean to reflect the solar radiation back into the atmosphere and prevent the ocean from absorbing it. They also include injecting chemicals like sulphur dioxide into a layer of our atmosphere called the stratosphere to prevent solar radiation from entering, making clouds closer to the ocean's surface more reflective, as well as thinning clouds that are higher up to help heat escape better. Of the previously mentioned processes, currently the favoured one is the chemical injection into the stratosphere, and I would agree it is the best choice as it is inexpensive, easy, and quick to do, and will have immediate effects.

Sulphur dioxide is naturally occurring and showed its strong cooling effects on the planet during the 1991 eruption of Mount Pinatubo, which sent thousands of tons of sulphur dioxide into the atmosphere, temporarily blocked out enough of the solar radiation to reduce global temperatures by one degree Fahrenheit. Injecting chemicals into the stratosphere might work well because it is very calm up there, and it has been estimated that gases can stay suspended there for six months to two years. This would buy us more time to work on better solutions for decreasing our GHG concentrations and production. On the negative side, the ozone layer is also in the stratosphere, and introducing chemicals like sulphur dioxide into the ozone layer could damage it and prevent it from recovering which means that when we decide to not use this option anymore, the planet's natural ability to reflect solar radiation and prevent harmful UV radiation from entering would be damaged, causing serious health and environmental issues. It is a good solution for a quick emergency fix, but it is not a long-term solution. Introducing sulphur dioxide into the atmosphere can also cause issues like acid rain, which is very destructive to the environment and human health.

The main reason I don't like the proposed techniques, which focus on reflecting the solar radiation, is that with all of them, there are too many unknown unknowns, and I believe we have tinkered with the chemical balance and our atmosphere enough. Also, we already have a major issue with microplastics in the ocean, and I really do not think we need more little pieces of inorganic material floating around in the ocean for aquatic life and seabirds to think is food, consume, eventually die from, and have it make its way up the food chain. Cloud thinning is a good idea as well, and I understand the thought behind it. Clouds trap a lot of heat close to the surface of the Earth, and if there were a way to thin the clouds, I can see how that would help cool down the surface of the Earth. Unfortunately, I have seen too many decisions being made because they are cheap and easy, only to have them fall apart because of the unknown unknowns occurring. Fixing the environment is something we need to do right the first time, and a key aspect to that is to know the risks and minimize or eliminate them from happening for the solution to be successful. Alteration of clouds, the weather, and adding additional chemicals all just seem like there is a very high risk of something unforeseen happening and making things worse. I believe the best/safest solution to reduce the amount of solar radiation from entering the atmosphere is with the construction of a physical barrier instead of a chemical solution.

The first possibility I am in favour of is building a temporary structure outside of the Earth's atmosphere to block a percentage of the solar radiation and prevent it from even getting into Earth's atmosphere. I know this sounds like a super villain's plan, but can you think of a better way to prevent the Sun's radiation from entering the Earth's atmosphere without messing around with the already struggling atmosphere? We would not need to create a structure that would block out the entire area of the Sun, but instead create a smaller structure that would be more mobile and could be moved during the different seasons to provide shade for regions that are struggling with heat-related issues. If you think that would require too much fabric, just keep in mind that approximately 400,000 km2 of fabric is made in one year, and I am sure that number can increase. I am sure that something like this would require a special material, but I am confident we would be able to make it work with some more research and development. If we can get the same commitment from governments and manufacturers as we did during war times and just focus most of our attention and resources on the goal of saving the planet for a couple of years, we could make this and other proposed projects in this book happen. Another issue is with getting the material into outer space, as it is very expensive to create and send a spaceship to outer space, as well as the amount of fossil fuel burned in the process would be very counterproductive.

Luckily, there is new technology that does not use any fossil fuels to launch objects into space; instead, it uses a vacuum-sealed centrifuge to spin an object like a satellite around very quickly and then launch it into space. It does require a large amount of energy, and we will discuss the energy production sector shortly. The main pros for the idea of building a temporary structure in outer space are:

1. It does not mess with the fragile chemical balance of the atmosphere.
2. It prevents solar radiation from even entering the atmosphere.
3. We can easily control how much solar radiation we let in by adding to it/dismantling it, moving it closer to or further away from the Earth, and moving it north or south compared to the Earth to help provide shade to areas experiencing heat-related issues.
4. It is temporary and we can easily dismantle it, send it off course, or destroy it when we don't need it anymore.
5. Low risk of unknown unknowns occurring as there is not a lot happening in the region of space just outside our planet besides a space station, space junk, and satellites flying around in a low orbit.

The cons include the following:

1. It will take years to build.
2. It will be expensive and require a lot of resources.
3. Public opinion will most likely be negative as they can't understand the urgency of the situation we are in and the need for something like this.

The cheap and dirty method I spoke of before also utilizes the centrifuge launcher, but instead of launching materials to build a temporary structure, we could launch and release something like garbage or small reflective pieces of metal into space and release it in a way that it remains in orbit around the Earth. If we send up enough material, we could create a barrier around the planet that significantly reduces the amount of solar radiation getting through, and it could also help with our waste management issue. Again, I know how this sounds, and I do feel silly saying it, but the pros are as follows:

1. It would help with two major issues we have with our environment: waste management and excessive solar energy entering our atmosphere.
2. Less expensive than other options.
3. Quick to execute with quick results.

Cons for this idea include:

1. Might cause issues with satellites and sending rockets to outer space.
2. Very difficult to remove the garbage when we have reduced the greenhouse concentrations to where they need to be.
3. Requires high amounts of energy to launch the pods.
4. No control

Are you still with me here? Again, I do realize how outlandish it sounds to suggest building a structure in outer space to block a portion of the Sun's radiation, or launch garbage into outer space. Still, unfortunately, we are at the point of extreme solutions for our extreme problems, and we have to get it right. To ensure we do it right, we need a solution that is simple and has a low risk of failure or causing issues from unknown unknowns; if you have any better ideas, please make them known.

Thawing Permafrost

A major issue with the solar radiation from the Sun heating up the atmosphere is the melting ice caps and thawing the permafrost. With the melting sea ice, the solar radiation is now also being absorbed by the oceans instead of being reflected back into the atmosphere and partially back into outer space, which is also directly increasing the ocean's temperature. Permafrost is estimated to cover 22.8 million km2 of the globe, and I found a great definition of permafrost in an article by Leslie Anthony with the Canadian National Geographic, titled, *"Arctic Permafrost is Thawing. Here's what it means for Canada and the World,"* from April 02, 2019, she describes permafrost as the following:

"Permafrost's simple definition — rock or soil that remains at or below °C for at least two consecutive years — is where the simplicity ends. Permafrost consists of an active layer of variable thickness that thaws and refreezes each year, and sits atop a deeper, more permanent layer. The distribution of permafrost on a landscape is recognized as continuous if it covers 91 to 100 per cent of the area, discontinuous if that coverage is 51 to 90 per cent, sporadic if it's 10 to 50 per cent, and isolated if it's less than 10 per cent (as with much of alpine permafrost). Permafrost can range from dry, with low water content, to ice-rich, where the ice content exceeds the saturated moisture capacity of the soil. "

We have briefly discussed the issue regarding how the permafrost was keeping the organic material frozen in the ground, preventing it from decomposing and releasing large amounts of methane into the atmosphere, adding to the greenhouse gas issue and how this would create a negative feedback loop; the permafrost melts, releases GHGs, makes the planet hotter, melts more of the permafrost, releases more GHGs, makes the world even hotter etc. With this in mind, we must develop a way to prevent the permafrost from thawing and give it a better chance to refreeze, which will prevent it from releasing GHGs into the atmosphere, stop the negative feedback loop, and prevent the Earth from passing its tipping point/point of no return.

If you look at what ice and snow do, due to their high albedo, they reflect much of the Sun's radiation and provide cover and insulation for the ground below. The large areas that were once covered by snow and ice are now bare and directly exposed to the Sun's radiation, and the previously frozen ground is now thawing. To prevent this from continuing, we need to implement a solution which will provide the same benefits as the snow and ice previously did, and we can accomplish this by creating an artificial cover over the thawing areas, first to provide shade to let its surface remain cool and refreeze, and second to do what the snow and ice was doing before and reflect the solar radiation back into the atmosphere.

The simplest and least expensive solution for this would be to install small structures, similar to solar panels, with a white or reflecting coating facing the Sun. The structures would be installed in a way that provides adequate shade for the ground below and is slanted enough so that it reflects the solar radiation back at different times of the year. These structures would not be constructed across the entire 22.8 million km2 of areas with permafrost, but just in areas where the ground has thawed or is starting to, the temperatures still hover around freezing, just to help give the Earth a little extra nudge to help refreeze while other global operations focusing on reducing greenhouse gasses and reducing incoming solar radiation are occurring. These areas do consist of forest, rock outcrops as well as large open spaces, and the idea would be to focus on the large open spaces which have thawed and are currently releasing methane. Another selling point regarding these structures is that they can be permanent or temporary. We will have the option to remove some or all of them at the point where we get the planet back to functioning as it should be, or they can be left in as they won't be disrupting anything there.

During the installation phase, local wildlife might be a bit spooked, and maybe some vegetation would be disturbed, but I am sure a thorough environmental assessment would determine that the benefits of installing this structure would outweigh the impact of the project. The shade from the structures would provide shelter for the smaller wildlife in the area and give them a better chance to scavenge for food while hiding from predators. The materials used for this could even be made from recycled material, which helps with the circular economy model, and since the design is so simple, it could be sourced from anywhere. Once complete, large-scale cloud seeding utilizing electrical charges could be used to increase the amount of snow cover and start regrowing the ice caps.

The local Indigenous groups would need to be consulted to ensure they approve of the environmental work happening on their lands, and they could also help provide workers to help with the installation and general maintenance of the structures post-installation.

Heat Islands

We discussed albedos previously in the permafrost section, but as a quick refresher, it's basically a measurement of something's ability to absorb or reflect the Sun's energy; darker surfaces absorb more energy while lighter colored surfaces reflect more energy. Just think, on a hot day, would you rather wear all white or all black? The material itself also plays a role in how much heat is absorbed, just think of it on a hot day, would you rather walk barefoot on a concrete or on a grassy field? Before humans started building large urban areas, massive cities, and roadways, most of the land was covered with vegetation like grass, bushes, and trees. Concrete, asphalt and other building materials absorb a lot more energy than vegetation, so what is happening now is more heat is being trapped at the surface of the Earth in the urban areas, and then slowly released when what should have happened is the heat being reflected away from the Earth's surface as it does in the surrounding regions of the large urban developments. Studies have shown that during the daytime, urban areas can reach temperatures 1-7 degrees hotter than the surrounding areas, while at night, it can be 2-5 degrees warmer. Thermal images of the Earth show massive amounts of heat being stored and radiating out of cities, while the surrounding areas of vegetation are not causing any concerns.

If you live in a large city, I am sure you can agree that in the summertime, you can feel the heat radiating off the streets, sidewalks, parking lots, and buildings, but if you were to go into a field or forest, it would be much cooler. By creating these Heat Islands where heat is being stored and slowly released instead of being reflected or quickly released, we are steadily adding heat close to the surface of the Earth and adding to the problem. It is also not just all the concrete and asphalt absorbing heat from the

Sun, but also the concentrated populations and the additional activities we do in these areas, that contribute to the increased temperatures in the urban areas. Transportation, commercial and industrial activities, and even how the city is designed, reducing wind flow, which could have cooled the city down, also add to the amount of heat the city holds in. Just as the previously mentioned major environmental concerns have negative feedback loops, these heat islands have their own.

With these high temperatures, to keep people comfortable and safe, many residential, commercial, and industrial facilities utilize air conditioning units and fans to help maintain a cooler temperature. Unfortunately, the additional demand for electricity means that the power companies need to produce more power to meet the demand, but many of the power generation facilities are still using fossil fuels to create energy. As we now know, burning fossil fuels creates additional GHGs, which increases the amount of heat stored in our atmosphere, making it hotter, increasing the amount of energy required to cool the insides of buildings, which then requires the energy producers to burn even more fossil fuels to meet the demand and pump more GHGs in the atmosphere etc.

There are building techniques which are now becoming mainstream to help create more sustainable buildings with programs like the "Leadership in Energy and Environmental Design" (LEED); they are promoting things like green roofs, rainwater collection, better natural interior climate control, more energy efficient processes as well as new methods which urban developers are using to design cities to prevent it or reduce its ability to become a heat island.

Seeing as demolishing and rebuilding entire cities is not an option, and since there is a very low chance that people living in the densely populated areas are all going to move out of the major cities into rural areas, become self-sustainable and environmentally sustainable, we must figure out ways to work with the current system. Simple changes like painting roofs and roads white would make a positive impact. The United States Environmental Protection Agency (EPA) proposes the following activities to reduce urban heat islands: increasing trees and vegetation cover around the city, installing "green" roofs, installing cool/reflective roofs, using cool pavements (either reflective or permeable), and utilizing smart growth technologies. Dr. Theresa Tam, Chief Public Health Officer of Canada, wrote an interesting article titled, "Reducing Urban Heat Islands to Protect Health in Canada," published April 29, 2020, and a link to it will be in the reference section in the back of the book. In the article, she discusses the causes and effects of the heat islands, but also case studies where local communities made simple changes like Toronto's Shade Policy and Guidelines, Improving thermal comfort in parks in Windsor, Ontario, Quebec City pilot project to green Anna Street, and Vancouver's Street Tree Cooling Networks.

Strategically adding trees and vegetation to your urban environment is an amazing way to help reduce the effects of urban heat islands. Plants naturally decrease the temperatures around them, not only by providing shade, but they also have two useful processes: evaporation and transpiration, together, they are called evapotranspiration. The *evaporation* aspect is from the rainwater being collected on the leaves and in the soil, and slowly being released in the surrounding area, cooling it down. *Transpiration* is similar, but it is the process of the tree absorbing water from the soil through its roots and then releasing it through its leaves, and the result of this is the cooling of the surrounding area. Research has shown that forested urban areas are on average 2.9 degrees Fahrenheit cooler than non-forested urban areas. The EPA has stated the following benefits of urban forests:

1. They create a lower energy use due to the lower demand for air conditioning.
2. Improve air quality and lower greenhouse gas emissions.
3. Enhances storm water management, water quality, and the soil and vegetation absorb and filter excess rainwater.
4. Reduce pavement maintenance since the shade that they provide can slow the deterioration of the pavement from the solar radiation.
5. Improve quality of life as vegetation can reduce noise, provide habitats for species, and cause a calming effect for people.

Additional cost savings for the municipalities have been discovered as well. A study was conducted in five U.S. cities, ranging in climate, and it was found that the cities spent roughly \$15-\$65/tree (including planting, pruning, and inspecting) and had a net annual benefit of \$30-\$90 per tree. The EPA has compiled a long list of useful information if you are interested in digging deeper into this; if you are interested in exploring this topic more, I highly recommend you read through their material to create a thorough understanding of the strategic implementation and benefits of increasing urban vegetation as there are many tips and tricks to maximize its efficiency.

A simple solution, which was used in one of the case studies mentioned by Dr. Tam, is removing a small number of parking spots on the street or expanding the sidewalks to create more room to plant vegetation. I know how frustrating it is to find parking in the city, but hopefully the increased vegetation will cool and clean the air enough to make the extra block or two of walking more relaxing and enjoyable for you. The decreased parking might also promote people to walk/bike more, take public transit more, or even create more of a reason for people to work at home instead of having to drive to work every day and struggle to find parking.

Any of these results will, of course, reduce the amount of GHGs from people driving everywhere and decrease their energy consumption from having to charge their electric vehicles.

Planting vegetation like grass and shrubs on the roof of your home or workplace to create a green roof can also play a major role in reducing the effects of heat islands. Just a side note, before adding additional weight to your roofs, please consult your friendly neighborhood civil or structural engineer, even a qualified building inspector might be able to ensure the roof is in good enough shape to support the additional weight. Green roofs need to be constructed properly to help with vegetation growth, water collection, proper drainage, and insulation. Green roofs, improperly constructed, can cause major damage to your roof, so please ensure you utilize professionals or conduct extensive research and use the appropriate materials and processes if you want to do it yourself. The benefits of green roofs include preventing the roof from absorbing the Sun's radiation and improving the insulation of the building (reducing the energy use by 0.7%). If used widely, green roofs can cool an urban area by up to 5 degrees F. They can also provide a habitat for wildlife as well as a nice green space for people to enjoy and unwind.

Green roofs are very cool, but "cool roofs" are also cool. "Cool roofs" have a high albedo and reflect more of the Sun's radiation than conventional roofing materials. The simplest way to improve your roof's albedo is to paint it a light color, use a more reflective material on it, or use a material that does not absorb as much heat energy as wood or terracotta shingles when compared to asphalt. Cool roofs reduce the temperature of the roof, inside the buildings as well as the surrounding ambient air and it does not absorb and slowly release as much heat as conventional roofs, consequently, buildings with cool roofs use less air conditioning, saving money by reducing the amount of energy used and potentially reducing the amount of GHGs from the energy production required to keep the inside of the building a comfortable temperature.

The final method we will discuss regarding reducing the heat island effects is changing how we build roads, sidewalks, and parking lots. This area is not as explored and developed as the previously mentioned areas, but seeing as 30-45% of the land cover in cities is covered by roads, sidewalks, and parking lots, it is something we need to look at. New processes to help improve these areas include making more reflective concrete and asphalt to help prevent them from absorbing so much heat. Other benefits to improving and utilizing the cool roads include:

1. Reducing stormwater runoff and improving water quality: The pavements can absorb more excess water and deliver it to the soil beneath the roads. The cooler pavement will also reduce the temperature of the runoff water and reduce the thermal shock to the aquatic life living in the areas where the runoff water flows into
2. Lower Tire Noise: The new pavement's ability to absorb water also reduces tire noise.
3. Enhanced safety: The pavements' ability to absorb water also decreases the amount of water pooling in the streets, causing vehicles to lose control.
4. Better nighttime visibility: The increased reflective pavement will improve visibility at night and potentially reduce the need for lighting requirements, reducing the amount of energy required.
5. Improve local comfort: Cool pavement will not store and radiate as much heat as conventional pavement, creating cooler spaces for people and wildlife to live.

Increasing Heat Waves, Heat Domes, Droughts, and Wildfires

Increasing heat waves have multiple negative effects on the environment and health. During heatwaves, the amount of energy required to meet the increased demand for air conditioning means that power generation facilities need to create more energy, which means that the facilities that use fossil fuels to create energy are creating more GHGs, which makes the climate warmer, thus the negative feedback loop previously mentioned in the heat island section. Heat waves also destroy agriculture, create droughts, and not only make it uncomfortable for everyone but can be fatal for those who have certain health issues. A head dome is when there is a high-pressure system that traps the hot air in a certain region and can be part of a heat wave; the high-pressure system, which creates a heat dome, is harmful as it prevents heat from leaving the area, even at night when it would just reflect the heat back down.

The intense heat destroys crops and limits the amount of water that can be used to provide the crops with the water they need to grow, limiting food supply. The extreme heat will also be extremely harmful to livestock, reducing their food and water availability, as well as creating environments too hot for them to survive. With the diminished water levels and stagnant water, we will see an increase in mosquito-transmitted diseases like West Nile due to the increasing populations of mosquitoes from the increasing stagnant bodies of water. The solution to decrease the heat waves and droughts is simple, but not easy: we need to cool the planet down by reducing the amount of solar radiation entering our atmosphere, reducing the amount of GHGs we produce, and removing the excess GHGs currently in our atmosphere. I have already discussed ways to reduce the amount of incoming solar radiation, and I will discuss ways to reduce the GHGs we are producing, as well as ways to reduce the amount of GHGs currently in our atmosphere.

The intense heat and drier weather bring an increasing number of wildfires. Wildfires have occurred for as long as there have been forests; it is natural for them to occur, and some are actually beneficial to the forest's health. As vegetation from trees and bushes die and collect on the forest floor, they pile up and prevent Sunlight from reaching the vegetation growing lower to the ground, preventing Sunlight from reaching the soil, and sometimes the tree's canopy can get very thick, which also prevents the Sunlight to come through and provide the required solar energy for plants to grown. Some of the wildfires help remove the dead and decaying material on the forest floor as well as open the forest canopy. The fires can also help release nutrients from the dead and decaying material, thereby improving the soil's fertility. Wildfires are especially beneficial for certain species of trees, plants, and flowers, as they require fire for their seeds to be released and germinate. To help with these processes, forestry experts conduct controlled burns to help control the amount of dead and decaying material on the forest floor and do their best to prevent it from piling up and creating large amounts of fuel to create large wildfires and help keep the forest healthy.

As the planet heats up, more vegetation is drying up and dying, and the amount of dead and decaying material in the forest is piling up faster than it can be dealt with. This generates large amounts of fuel to create the out-of-control wildfires, which we have been witnessing over the past couple of years. As I'm sure you know, when trees burn, they create smoke, and that smoke contains GHGs as well as many other harmful things like particulate matter, which is bad for both the environment and human health. Again, we have another negative feedback loop with the warmer weather creating drier forests, which create bigger fires, which create more GHGs, which make the world hotter, etc. During the smaller/controlled burns, the amount of material burned and smoke created is intentional, and causes negligible harm to the environment, but it is the massive out-of-control wildfires that cause serious damage to the environment, people's health, and their property. According to the Joint Economic Committee of Democrats, wildfires have caused between $394 to $893 billion dollars in economic costs and damages each year in the U.S.; these costs include diminished real estate values, lost income, damage to watersheds and aquifers, insurance payouts, property and infrastructure damage, electricity cost, evacuation costs, federal wildfire protection costs, school and leaning losses, insurance premium increases and tourism losses. In addition to that, there are health costs such as death and injuries from the fire and smoke, as well as psychological issues.

The main solution for this would be to improve forest management, which the U.S. government is working on, and I am hoping other countries will follow suit. It is an expensive and time-consuming process, but if we can get into the forests and help control the buildup of fuel in the form of dead materials on the forest floors, we will be able to reduce the size of the wildfires. New technologies can be developed and implemented as well, especially in the forested areas surrounding cities and farmlands.

Early detection of wildfires is very important as well. Technology like drones or satellites monitoring the forests would make a big difference, but it would be very expensive. A less expensive option would be for recreational pilots and pilots trying to get their hours to get together and set up a volunteer wildfire spotting campaign.

Regarding technologies that would be implemented to put out wildfires, one option could be something like large portable silos or tanks filled with fire suppression chemicals. These silos could be positioned in the forests surrounding urban areas or deeper in the forest to either be controlled remotely or manually accessed to release the fire suppressive chemicals and prevent the fire from spreading into the rural areas or the forest; this is a proactive approach instead of the current reactive approach. Other technologies which could be explored are projectiles that could be launched and detonated over a wildfire, releasing a large amount of fire suppressant chemical over a large area instead of having a plane drop it in a straight line. Imagine launching 10 projectiles and having them detonate over a large fire; they would be able to cover the area with a lot more fire-suppressing material than a fleet of planes could do in a day. This may be more expensive, but it is more efficient. Looking at the annual amount governments spend on wildfires per year shows that governments would be interested in purchasing new technologies to save them from financial losses. Of course, reducing the temperature of the planet will play a major role as well, but until that happens, we must also focus on dealing with the symptoms of climate change.

Droughts are becoming more abundant due to the increasing heat and changing weather patterns. The Centers for Disease Control states that droughts can cause the following long-term public health problems:

1. Shortages of drinking water and poor quality of drinking water
2. Impacts on air quality, sanitation and hygiene, and food and nutrition
3. More diseases, such as West Nile Virus, are carried by mosquitoes breeding in stagnant water.

During droughts, stream and river flows will reduce and can increase the concentration of pollutants in the water. Higher water temperatures in lakes and reservoirs can negatively affect fish and other aquatic life, as well as overall water quality, by reducing oxygen levels. The ultra-dry soil can also cause flash floods and heavy runoff since the ground will not be able to absorb the water. There are many communities who also rely on well water to drink and use, but with the decreasing rainfall and the ground's inability to absorb water, these underground reservoirs would be depleted much sooner than anticipated, and people will be in serious trouble. Air quality is negatively affected during a drought as it creates dry and dusty environments. Toxins from bacteria growing in stagnant water can cause serious respiratory illnesses in certain populations. Reduced access to water also means reduced hygiene and sanitation methods, from personal hygiene to washing food. With a decrease in hygiene comes an increase in poor health. According to UNICEF, the decreased amounts of water can also lead to a failure in the sewage systems, and the threat of contracting diseases like cholera surges. Some facts from UNICEF regarding water scarcity include:

1. Four billion people, almost two-thirds of the world's population, experience severe water scarcity for at least one month each year.
2. Over two billion people live in countries where the water supply is inadequate.
3. Half of the world's population could be living in areas facing water scarcity by as early as 2025.
4. Some 700 million people could be displaced by intense water scarcity by 2030.
5. By 2040, roughly 1 in 4 children worldwide will be living in areas of extremely high water stress.

In some areas around the world where water is scarce, it is mainly up to the women and children to travel long distances to collect water, which means less time at school and exposes them to additional safety risks, exploitation, and an intense physical burden on their lives.

Operations are underway to create drinking water from ocean water by removing the salt from it; this process is called desalination. There are two main processes by which the salt is removed from the ocean water. The main one is using reverse osmosis, where the water moves through an electrically charged membrane, which removes the salt. The second is through heating it up, collecting the steam, then cooling the steam down so it turns into clean water. The major issue with these processes is that they both produce wastewater, which has a high salt concentration (brine), which ends up being dumped back into the ocean, causing a toxic environment for aquatic life, or disposed of on land. Producers are trying different methods to dilute the brine as it is pumped into the ocean, as studies have found that it can change the composition of species living in the discharge area and create a hazardous environment for others. Another issue, specifically with the one involving heating the water up, is that the waste material has a much higher temperature than the water where it is discharged, and can create an inhabitable area.

Too Many GHGs

This section will explore the major contributors of human-made GHGs and provide insight on what we need to do to reduce their production. In 2021, 54.6 billion tons of GHGs (carbon dioxide equivalent) were generated, and scientists are estimating that the Earth is removing approximately 2 billion tons of GHGs per year. Natural carbon sinks are still the main reason for the carbon reduction, but as previously noted, the forests are shrinking, and the ocean's ability to capture carbon is reducing as it is becoming more acidic and killing some of the aquatic life that has previously been capturing the carbon for us. Billions of dollars are being spent on developing carbon-capturing technology, but the best remedy is to leverage our natural resources. The Paris Agreement is an agreement that was created by the UN to encourage governments to create a plan to reduce their GHG emissions, remove GHGs, and then execute their plan. Its goals are as follows:

The Agreement sets long-term goals to guide all nations to:

- Substantially reduce global greenhouse gas emissions to hold global temperature increase to well below 2°C above pre-industrial levels and pursue efforts to limit it to 1.5°C above pre-industrial levels, recognizing that this would significantly reduce the risks and impacts of climate change
- Periodically assess the collective progress towards achieving the purpose of this agreement and its long-term goals
- Provide financing to developing countries to mitigate climate change, strengthen resilience, and enhance abilities to adapt to climate impacts.

On November 4, 2016, the legally binding (but not enforceable) Paris Agreement came into action with almost every nation signing onto it. The Accord's main milestones happen every five years. In 2020, countries submitted their plans, and starting this year, due in 2025, countries will start reporting on the progress of their plans. As there is a chance that the Paris Agreement will not meet its goals, we also need to rely on the private sector to help fix the planet.

The total U.S. Greenhouse Gas Emissions by Economic Sector are as follows (CO2 Equivalent):

Transportation (28%): This is the highest greenhouse gas creator in the U.S. and comes from burning fossil fuels to power our cars, trucks, ships, trains, and airplanes.

Electrical Power (25%): Power generation is the second biggest producer of GHGs. In 2021, in the U.S., approximately 60% of the electricity came mainly from burning coal, natural gas, and a small percentage of petroleum.

Industrial (23%): The industrial sector produces the goods and materials we use every day, and there are two categories of emissions that the EPA took into consideration: direct emissions and indirect emissions.

Commercial and Residential (13%): This area was the fourth-largest greenhouse gas producer in the U.S. in 2021. GHGs come from burning fossil fuels for heating and cooking, waste management, and the use of gas for cooling.

Agriculture (10%): The GHGs generated from agricultural activities are from livestock such as cows, agricultural soils, and rice production. The livestock produces methane from its digestive system, and nitrous oxide is created from the nitrogen in the fertilizer.

Now let's look at another major generator of GHGs, like China, to compare where their greenhouse gases are coming from. According to the IEA, the leading sources of China's GHGs are as follows: Power Generation (48%), Industry (36%), Transport (8%), and Residential and Commercial (5%).

Comparing numbers between countries shows that each country needs to make its own specific plan to reduce its GHG production. Since China has such a large manufacturing sector and population, it makes sense that their largest greenhouse gas-producing industries are power generation and industrial, while the largest greenhouse gas-producing industries in the U.S. are transportation and power generation. Looking at these statistics, it shows that power generation is a major contributor to both countries. An interesting point to add is that the power generation sector's greenhouse gas emissions will increase globally as more electric cars are being used and charged in areas that use fossil fuels for electricity.

The table below provides examples from the EPA for reduction opportunities for each of the main sectors:

Power Generation

Type	How Emissions Are Reduced	Examples
Increased Efficiency of Fossil-fired Power Plants and Fuel Switching	Increasing the efficiency of existing fossil fuel-fired power plants by using advanced technologies, substituting less carbon-intensive fuels, and shifting generation from higher-emitting to lower-emitting power plants	• Converting a coal-fired boiler to use natural gas, or co-firing natural gas. • Converting a single-cycle gas turbine into a combined-cycle turbine. • Shifting the dispatch of electric generators to lower-emitting units or power plants.
Renewable Energy	Using renewable energy sources rather than fossil fuels to generate electricity.	Increasing the share of total electricity generated from wind, solar, hydro, and geothermal sources, as well as certain biofuel sources, through the addition of new renewable energy generating capacity

Increased End-Use Energy Efficiency	Reducing electricity use and peak demand by increasing energy efficiency and conservation in homes, businesses, and industry.	EPA's ENERGY STAR® partners avoided over 400 million metric tons of GHGs in 2020 alone, helped Americans save over $42 billion in energy costs, and reduced electricity use by 520 billion kWh.
Nuclear Energy	Generating electricity from nuclear energy rather than the combustion of fossil fuels.	Extending the life of existing nuclear plants and building new nuclear generating capacity.
Carbon Capture and Sequestration (CCS)	Capturing CO2 as a byproduct of fossil fuel combustion before it enters the atmosphere, transporting the CO2, and injecting the CO2 deep underground at a carefully selected and suitable subsurface geologic formation where it is securely stored.	Capturing CO2 from the stack of a coal-fired power plant and then transferring the CO2 via pipeline, injecting the CO2 deep underground at a carefully selected and suitable nearby abandoned oil field, where it is securely stored.

Transportation

Type	How Emissions Are Reduced	Examples
Fuel Switching	Using fuels that emit less CO_2 than the fuels currently being used. Alternative sources can include biofuels, hydrogen, electricity from renewable sources, such as wind and solar, or fossil fuels that are less CO_2-intensive than the fuels that they replace. Learn more about Green Vehicles and Alternative and Renewable Fuels.	•Using public buses that are powered by renewable fuels, electricity, or compressed natural gas rather than conventional gasoline or diesel. • Using electric, plug-in hybrid electric, or hydrogen fuel cell vehicles. • Using renewable fuels such as low-carbon biofuels.
Improving Fuel Efficiency with Advanced Design, Materials, and Technologies	Using advanced technologies, design, and materials to develop more fuel-efficient vehicles. Learn about EPA's vehicle greenhouse gas rules.	• Developing advanced vehicle technologies such as hybrid vehicles and electric vehicles, that can store energy from braking and use it for power later. • Reducing the weight of materials

		used to build vehicles. • Reducing the aerodynamic resistance of vehicles through better shape design.
Improving Operating Practices	Adopting practices that minimize fuel use. Improving driving practices and vehicle maintenance. Learn about how the freight transportation industry can reduce emissions through EPA's SmartWay Program.	• Reducing the average taxi time for aircraft. • Driving sensibly (avoiding rapid acceleration and braking, observing the speed limit). • Reducing engine-idling. • Improved voyage planning for ships, such as through improved weather routing, to increase fuel efficiency.

Reducing Travel Demand	Employing urban planning to reduce the number of miles that people drive each day. Reducing the need for driving through travel efficiency measures such as commuter, biking, and pedestrian programs. Learn about EPA's Smart Growth Program.	• Building public transportation, sidewalks, and bike paths to increase lower-emission transportation choices. • Zoning for mixed use areas, so that residences, schools, stores, and businesses are close together, reducing the need for driving.

Industry Sector

Type	How Emissions Are Reduced	Examples
Energy Efficiency	Upgrading to more efficient industrial technology. EPA's ENERGY STAR® program helps industries become more energy-efficient.	Identifying ways that manufacturers can use less energy for industrial processes and to run equipment.
Fuel Switching	Switching to fuels that result in less CO_2 emissions but the same amount of energy, when combusted.	Using natural gas instead of coal as process input.
Recycling	Producing industrial products from materials that are recycled or renewable, rather than producing new products from raw materials.	Using scrap steel and scrap aluminum as opposed to smelting new aluminum or forging new steel.
Training and Awareness	Making companies and workers aware of the steps to reduce or prevent emissions leaks from equipment. EPA has a variety of resources for training and other steps for reducing emissions. EPA has	Instituting handling policies and procedures for perfluorocarbons (PFCs), hydrofluorocarbons (HFCs), and sulfur hexafluoride (SF_6) that reduce occurrences of accidental

	experience working with the aluminum, semiconductor, and magnesium industries.	releases and leaks from containers and equipment.

Residential and Commercial Sector

Type	How Emissions Are Reduced	Examples
Homes and Commercial Buildings	Reducing energy use through energy efficiency.	Homes and commercial buildings use large amounts of energy for heating, cooling, lighting, and other functions. Energy-efficient building practices and retrofits can allow new and existing buildings to use less energy to accomplish the same functions, leading to fewer greenhouse gas emissions. Techniques to improve building energy efficiency include better insulation and building envelope improvements; more energy-efficient heating, cooling, ventilation, and refrigeration systems; efficient LED lighting; passive heating and lighting to take advantage of

		Sunlight; and the purchase of energy-efficient appliances and electronics. Learn more about ENERGY STAR®.
Wastewater Treatment	Making water and wastewater systems more energy-efficient.	Drinking water and wastewater systems account for approximately 2% of energy use in the United States. By incorporating energy efficiency practices into their water and wastewater plant, municipalities and utilities can save 15 to 30% in energy use. Learn more about Energy Efficiency for Water and Wastewater Utilities.
Waste Management	Reducing solid waste sent to landfills. Capturing and using methane produced in current landfills.	When solid waste decomposes in landfills, it creates landfill gas, which is primarily comprised of CO2 and CH4. There are several well established, low-cost methods to reduce GHGs from consumer waste,

		including recycling programs, waste reduction programs, and landfill methane capture programs. • Learn about recycling • Learn about WARM, EPA's Waste Reduction Model. • Learn about EPA's Landfill Methane Outreach Program, which promotes the recovery and use of landfill gas. • Learn about appliance recycling from EPA's Responsible Appliance Disposal program.
Air Conditioning and Refrigeration	Reducing leakage from air conditioning and refrigeration equipment. Using refrigerants with lower global warming potentials.	Commonly used refrigerants in homes and businesses include ozone-depleting refrigerants and blends consisting entirely or primarily

		of hydrofluorocarbons (HFCs). Both HCFCs and HFCs are potent GHGs. In recent years, there have been several advancements in air conditioning and refrigeration technology that can help homes and businesses reduce both refrigerant charges and refrigerant emissions. For instance, in the retail food sector, learn more about EPA's GreenChill Program to reduce greenhouse gas emissions from supermarkets.

Agriculture Sector

Land and Crop Management	Adjusting the methods for managing land and growing crops.	• Fertilizing crops with the appropriate amount of nitrogen required for optimal crop production, since over-application of nitrogen can lead to higher nitrous oxide emissions without enhancing crop production. • Draining water from wetland rice soils during the growing season to reduce methane emissions.
Livestock Management	Adjusting feeding practices and other management methods to reduce the amount of methane resulting from enteric fermentation.	• Improving pasture quality to increase animal productivity, which can reduce the amount of methane emitted per unit of animal product. Also, increased productivity in livestock can be introduced through improved breeding practices.
Manure Management	• Controlling the way in which manure decomposes to reduce nitrous oxide and methane emissions.	• Handling manure as a solid or depositing it on pasture rather than storing it in a liquid-based system, such as a lagoon, would likely reduce methane emissions but may

	•Capturing methane from manure decomposition to produce renewable energy.	increase nitrous oxide emissions. • Storing manure in anaerobic lagoons to maximize methane production and then capturing the methane to use as an energy substitute for fossil fuels. • For more information on capturing methane from manure management systems, see EPA's AgSTAR Program, a voluntary outreach and education program that promotes the recovery and use of methane from animal manure.

"Our World In Data" has created an interactive chart which shows the global greenhouse gas emissions per sector in carbon dioxide equivalent over the past 30 years, and I will add a link to it in the sources. The chart shows how power generation and heating contribute the most to greenhouse gas emissions globally. In 2020, the top four global emission generators by industry are as follows: Electricity and Heat (15.8 billion tons), Transportation (7.29 billion tons), Manufacturing and Construction (6.22 billion tons), and Agriculture (5.87 billion tons), totaling approximately 50 billion tons.

Every country has different industries producing different amounts of GHGs. If the country is a large producer of manufactured goods to meet the world's demand, it would be producing a much higher amount of GHGs from power generation required for their factories to function when compared to a country whose main focus is on agriculture to meet the world's demand. The Paris Agreement is useful in the sense that it asks each country to investigate their individual activities and requirements and make a customized plan that will meet the GHG reduction goals.

Power Generation

There is currently an initiative called Net Zero Emissions (NZE) by 2050, and it is essentially a pathway for the energy sector to achieve a neutral amount of CO2 emissions by 2050; that means the energy sector does not create any CO2, or it removes the same amount that it created.According to the International Energy Agency (EIA), the initiative includes the following:

- Describes a pathway for the global energy sector to reach netzero CO_2 emissions by 2050 by deploying a wide portfolio of clean energy technologies, without offsets from land-use measures.Decisions about technology deployment are driven by costs, technology maturity, market conditions, available infrastructure, and policy preferences.
- Prioritizes an orderly transition that aims to safeguard energy security through strong and coordinated policies and incentives that enable all actors to anticipate the rapid changes required, and to minimize energy market volatility and stranded assets.Rapid deployment of clean energy technologies and energy efficiency is at the core

of this transition. The NZE Scenario is underpinned by a detailed analysis of project lead times for minerals supplies and clean energy technologies as part of efforts to ensure the feasibility of the deployment. There is, however, inevitably a risk of bottlenecks emerging for some technologies, which underscores the importance of measures to enhance material reuse and recycling and to drive down the material intensity of clean energy technologies.

- Recognizes that achieving net zero energy sector CO_2 emissions by 2050 depends on fair and effective global co-operation. The pathway to net zero emissions by 2050 is very narrow. All countries will need to contribute to deliver the desired outcomes; advanced economies take the lead and reach net zero emissions earlier in the NZE Scenario than emerging market and developing economies. Global access to electricity and clean cooking is achieved by 2030 in line with established SDGs. Rapid and major reductions in methane emissions from the oil, gas, and coal sectors help to buy some time for less abrupt CO_2 reductions in emerging markets and developing economies. Global collaboration facilitates the development and adoption of ambitious policies, drives down clean technology costs, and scales up diverse and resilient global supply chains for critical minerals and clean energy technologies. Enhanced financial support to emerging markets and developing economies plays a critical part in this collaboration.

As the population grows and we become more dependent on technology for work, leisure, daily transportation, traveling, and manufacturing, the need for more power generation will increase.This has been noted, and many advancements in power generation are currently being researched, have been developed and are currently being implemented, or are already in use and are being improved upon.The demand for energy creation will also significantly increase over the next decade as there are plans to provide everyone in the world with access to electricity to help improve their lives and significantly decrease the amount of fossil fuels used in developing countries for daily requirements like cooking and heating, which will also significantly improve air quality in the regions.Below is a table to show the goals of the global electrification plan by 2030:

Access to electricity		**Access to clean cooking**	
Universal access by 2030	**Other targets**	**Universal access by 2030**	**Other targets**
	Africa		
Angola, Benin, Botswana, Burundi, Cabo Verde,	Burkina Faso, Central African	Angola, Kingdom of Eswatini, Kenya,	Benin, Botswana, Burkina Faso,

Access to electricity		Access to clean cooking	
Universal access by 2030	**Other targets**	**Universal access by 2030**	**Other targets**
Cameroon, Côte d'Ivoire, Eritrea, Ethiopia, Gambia, Ghana, Guinea, Kenya, Liberia, Mauritania, Mozambique, Nigeria, Rwanda, Senegal, United Republic of Tanzania	Republic, Chad, Republic of the Congo (Congo), Djibouti, Democratic Republic of the Congo (DRC), Equatorial Guinea, Guinea-	Malawi, Mali, Mozambique, Rwanda, Sudan	Burundi, Cabo Verde, Cameroon, Côte d' Ivoire, DRC, Ethiopia, Ghana, Guinea, Guinea-Bissau, Liberia, Madagascar, Mauritania,

Access to electricity		Access to clean cooking	
Universal access by 2030	Other targets	Universal access by 2030	Other targets
(Tanzania), Togo, Zambia	Bissau, Lesotho, Madagascar, Malawi, Mali, Namibia, Niger, Sierra Leone, Somalia, South Sudan, Sudan,		Niger, Nigeria, Senegal, Sierra Leone, Tanzania, Togo, Uganda, Zimbabwe

Access to electricity		Access to clean cooking	
Universal access by 2030	**Other targets**	**Universal access by 2030**	**Other targets**
	Uganda, Zimbabwe		
	Developing Asia		
Myanmar, Nepal, Pakistan	Cambodia, Papua New Guinea	Bangladesh, China, Fiji, India, Indonesia, Mongolia, Myanmar,	Bhutan, Pakistan, Philippines, Thailand

Access to electricity		Access to clean cooking	
Universal access by 2030	**Other targets**	**Universal access by 2030**	**Other targets**
		Nepal, Viet Nam	
	Central and South America		
Bolivia, Dominican Republic, Ecuador, El Salvador, Guatemala, Honduras,	Haiti	Colombia, Peru	Haiti, Honduras, Guatemala, Mexico, Nicaragua

Access to electricity		Access to clean cooking	
Universal access by 2030	**Other targets**	**Universal access by 2030**	**Other targets**
Nicaragua, Panamá, Peru, Trinidad and Tobago			

Clean energy sources will need to supply the power to these developing areas. Luckily, there have been some amazing advancements in creating clean energy as well as energy storage.

Advancements in Clean Energy Technology

The first power generation facilities used the energy from falling water or coal to generate their power, since then, we have made advancements in clean energy generation which also captures the energy from falling water but also the ocean's movement, the Sun's energy, wind's driving power, as well as utilizing naturally existing heat from beneath the surface of the Earth. If you are one of those people who are not a fan of Nuclear Power, I will discuss it in more detail shortly, and hopefully, you will change your mind and see how beneficial and low-risk it can be.

Hydro

The kinetic energy from falling water has been useful to humanity for a very long time, and for many activities other than providing electricity. Currently, almost all hydroelectricity is generated in hydroelectric dams. Basically, the water level in a river or lake is increased by building a big wall downstream called a dam, which creates a reservoir. The dam then releases controlled amounts of water primarily to meet the energy demand of electricity and water levels, the flowing / falling water then spins a turbine, which spins copper wiring inside a magnetic field and creates energy. This form of energy creation is very clean, it does not give off any greenhouse gas emissions, is reliable, and controllable. Factors that can affect the hydroelectric power generation include the amount of precipitation in the region, how quickly the reservoir is evaporating due to the ambient temperature, environmental issues, and landslides. The locations of hydroelectric dams must be carefully chosen as they can create very negative effects on their environment due to their construction and changes made both upstream and downstream of the dam.

A newer source of energy, which utilizes the ocean's natural movements, is being developed. Most of the time, the ocean is very restless. It not only has small surface waves that you can easily see when you look out onto the ocean, but the ocean also shifts on a global scale four times a day, which is essentially the moon's gravity pulling at the ocean's water, creating a massive wave, commonly called a tide. The ocean is constantly moving in all directions due to waves and currents. Researchers are developing technologies that use the constant movement of the ocean to create energy. This technology will enable mankind to obtain power from a never-ending, reliable source. Unlike hydroelectric, global warming will essentially increase the amount of ocean water and make the ocean more violent, which in turn could create more energy. Finally, a positive feedback loop from climate change! With that being said, the ocean currents are actually slowing down due to the melting ice caps. Some of these new power generation processes, which create power from waves, tides, and currents, incorporate having the moving water spin a turbine, or use floating objects (called buoys) to move up and down, moving copper wire inside a magnetic field and creating energy. Many countries are using these technologies on a small scale as the technology is still being developed, but many of the countries can see related technologies being used in large amounts within the next 10 years. This process is also very clean as it does not give off any GHGs; the only downfall is that it might disrupt the local aquatic ecosystem. The locations of the ocean movement energy generation facilities will need to be specific.

Geothermal

Deep under our feet, beneath the Earth's surface is a lot of heat stored in the rocks, temperatures can increase by 72-87 degrees Fahrenheit per mile. The heat is generated from a couple of different things constantly happening beneath our feet; the breakdown of radioactive material releasing heat, and heat which is being generated in the core of the Earth, moving through the area between the core and the surface of the Earth (the mantle), and getting close to the surface. The heat generated by these processes is constant, but the depth at which they can be accessed will vary depending on where you are on the Earth's crust. Basically, geothermal power generation facilities work by drilling into the Earth's crust and tapping into large underground reservoirs of extremely hot water. The hot water and steam are then extracted and used to spin a turbine and generate energy. Like hydroelectric power generation previously mentioned, there are no GHGs being generated during this process, and its location needs to be specific. Geothermal is more reliable than power that relies on the weather, like solar and wind, and studies on geothermal claim that it can produce more electricity over the same time period than coal, natural gas, nuclear, or large hydro stations.

Solar

I am pleased to share that the efficiency of solar panels is greatly increasing while installation costs are decreasing. Previously, solar panels were only able to capture a fraction of the light generated by the Sun due to the single silicone layer on the panels, and then stored in batteries to be used on demand. Recently, technological breakthroughs have improved the efficiency of solar panels to increase the recovery of the Sun's energy from 24% to over 30%. Technology in energy storage (batteries) is improving as well. The breakthrough in increasing the efficiency of solar panels was from adding a second layer of a material called perovskite, which captures the blue light in the Sun's radiation while the silicone layer captures the red light. One thing to note that many people forget is that solar panels use the Sun's light, not the Sun's heat, to generate electricity, so the solar panels are still effective in cold regions; you would just have to clean the snow off. The surfaces of the solar panels can get very hot due to absorbing high amounts of solar radiation, and if we could recover the heat captured in these panels and utilize it for our benefit, we would be able to maximize their efficiency significantly. The amount of light from the Sun varies depending on where you are on Earth. Being closer to the equator and at a higher elevation like in the mountains will be the best location to have the highest intensity of solar energy and ability to recover the most power, while the further away from the equator and closer to ocean level you are, the less intense the Sun's energy will be the less power will be generated. This form of energy is also unreliable as it is dependent on clear skies (energy can still be created on cloudy days, just not as much), but if improvements to the panels' efficiencies continue to improve, the panels will be more useful in more areas instead of being used only in areas of abundant Sunshine.

Another suggestion for farmers is that if you have a lot of land that is not being used for grazing or crops, think about installing solar panels on it and selling the excess power back to your local energy provider. Currently, over 40 states in the U.S. and many areas in Canada allow "net metering," which means you are able to sell excess power generated on your property back to the power grid; there are also tax credits you can obtain from installing solar panels. Not only would this generate more money for the landowners in the long run, but it would also significantly help with the current energy crisis by producing more clean energy. Just think of all the empty unused spaces which could be generating income for the landowner along with clean power for the people, seems like a win-win to me. Solar panels can be expensive to install. Perhaps the government can provide grants to help farmers cover the cost of installing solar farms on their property, in fields that they cannot farm due to poor soil conditions, or if the fields are just getting too much for them to handle. This would benefit both the farmers and the environment.

Wind

Wind turbines are probably one of the most controversial forms of clean energy, other than nuclear, but advancements are being made to improve the generating power from wind. Traditionally, wind power is generated by the wind spinning a turbine to generate electricity. These turbines are generally very tall and usually have three blades on them and look like big tabletop fans. Issues which come from these turbines include the amount of noise they generate, negatively alter the appearance of the landscape, and can be very harmful to wildlife like birds and bats. New technological advances have been made in the wind power field, and one of them, developed by a company called Airloom, completely changed the design of wind power generating structures.

Instead of three large blades on a single tower, 82-foot-high poles hold up a track with 33-foot-long "wings" that are attached vertically. As the wind blows, the wings travel around the track and generate power. "Instead of flying in a circle, that blade is flying across the track and generating a mechanical force just like the blade of a wind turbine turns a gearbox in the center," says Neal Rickner, who recently joined the company as CEO.

Just like the previously mentioned green power-generating methods, wind power generation does not produce GHGs, but it is not as reliable as other methods, as wind can be unpredictable. The locations of the wind farms need to be specific as they need to be installed in areas that have high winds.

Nuclear

Nuclear power is great, it is reliable, simple, and efficient. Currently, there are 54 commercially operating nuclear power plants in the U.S., six nuclear power plants in Canada, and about 440 nuclear power reactors providing approximately 10% of the world's power in 2021; the other main sources of power in the world in 2021 are coal (36%), gas (23%), hydro (16%); solar, wind, tidal and geothermal add up to about 10.5%.

Basically, how energy is generated from nuclear material is that the nuclear material slowly breaks down through the splitting of atoms (called fission), a lot of heat is released. The heat is then used to create steam, which is then pressurized and used to spin a turbine; this process produces a large amount of energy consistently. The process does not create any GHGs, and we have the technology and controls in place to eliminate the exposure of radioactive energy from the radioactive material. Historically, nuclear power has suffered from serious incidents, but lessons have been learned; currently, we have been building nuclear power generating facilities, which currently provide 10% of the global power without you probably even knowing. The serious incidents from nuclear power facilities are from building nuclear power plants in Tsunami and earthquake-prone zones as well as human errors; however, the location of the nuclear facility needs to be properly sited. Other issues with nuclear power plants are that radioactive material required for the process is not a renewable resource, nuclear power plants are expensive and take a long time to build, they produce nuclear waste, and if they are built in the wrong area and were to break down, it can be devastating to the surrounding environment. If done properly, nuclear power could be the best chance we have at safely meeting the growing demand for clean electricity. In my opinion, the pros, including the amount of reliable, clean energy produced from nuclear power plants, outweigh the cons. Currently, we need to get away from using fossil fuels to create energy and focus on clean energy generation. All the clean energy generation options must be set up in specific areas due to their requirements; hydro needs to be set up on large rivers, solar needs to be setup in areas of abundant Sunshine, wind needs to be set up in windy areas, and tidal needs to be setup in the oceans. Nuclear facilities should be setup away from waterbodies, environmentally sensitive areas, and in areas with low risk of natural disasters. The fuel they use provides consistent energy, it is reliable and highly productive compared to other clean energy processes. With the increasing population and use of technology, the demand for power is increasing, and in order to meet it and not destroy our planet in the process, nuclear power generating facilities are currently our best hope.

Power Grid

You can't talk about power generation without talking about the power grid. Simply put, the power grid is just the roadways for the electricity to get from where the energy is being produced to the end user. In North America, Canada, the U.S., and Mexico are all on the same power grid and are able to move electricity (bought and sold) across borders to where the demand is. This can be good or bad depending on how the power is generated; even if you are living in an area that produces "green energy" like hydroelectric, you might still be using dirty energy from fossil fuels.

On a micro level, living "off the grid" is a concept whose literal meaning is living off the power grid and is self-sustainable, which means all the requirements like heating, cooling, electricity for cooking, etc. are dealt with independently, and not using any outside-generated electricity. Depending on how the living environment is set up, it could be the ideal way for people who live remotely to impact the environment, or it could be very harmful. Micro dams or even putting turbines directly in a river can power single homes or even a small community; if doing this, please obtain all permits as there could be major issues downstream regarding habitat destruction, changing the course of the river, and more. I frequently go to a remote resort which is off the grid, and they use small-scale hydro and solar to heat their water (showering, cleaning, etc.), power their lights and outlets to charge their laptops and phones so they can still work and live normal lives, and propane to cook their food. Good forestry practices are observed, and the dead wood on the forest floor is used in the occasional bonfire to help with forest management in their area; the fires are also only made during times of low wildfire risks. They may have gas generators on-site, but they are rarely used except in emergencies. Currently in British Columbia, Canada, many communities are still powered by massive diesel generators since it was too expensive to extend the power lines (power grid) to these remote areas; areas like this include Bella Coola, Bella Bella, Haida Gwaii, and Atlin. Living off the grid can be beneficial for the environment if done properly. There is also the waste management aspect of it (sewage, garbage, etc.), which will be discussed later, but regarding the environmental impact from energy production, it can be done properly or be highly destructive.

Energy Storage / Transportation

Research and development into storing energy has come a long way, and increasing the amount of energy we could store could really make a positive impact on how we operate. By saving excess power created during times of low energy demand (like at night when everyone is sleeping) and then using it to help support the times when energy demand is at its highest, it will help stabilize the demand for power and help reduce the excess use of dirty energy. One idea I have been exploring is having large container ships containing massive energy storage technology like batteries. These ships could have their batteries charged at large power storage facilities at ports in areas like the Middle East, Africa or South America, which are charged by clean energy such as solar farms. Once the batteries are charged, the stored energy is then transported to areas around the world that are struggling to have power. The power stored in the batteries on the ships can then be transferred to large energy storage facilities on shore at the ports or pumped directly onto the power grid to provide electricity to the rural areas. Currently, it's probably much cheaper to just build power-generating facilities in these isolated areas or just hook them up to a pre-existing power grid.

Transportation

Globally, transportation is the second-largest contributor of greenhouse gas emissions in our atmosphere, next to energy production. It produced about half of the greenhouse gas emissions from energy production globally in 2021. Transportation includes how you get to school, work, leisure activities, etc., as well as moving all the food, clothing, technology, and other consumer goods from international origins, crossing oceans and countries, or just domestically, within your own country. According to the International Energy Agency (IEA), in 2018. Transportation was responsible for one-fifth of the world's CO2 emissions, 45.1% from cars and buses, 29.4% from trucks carrying freight, 11.6% due to aviation, and 10.6% related to international shipping.

I am sure you have heard that you should try to walk, bike, take public transit, or even carpool instead of taking your vehicle to work, school, or to run errands. Things like walking, biking, and even taking public transportation are not just good for the environment, but also good for your health. I understand that many people live in areas where it's not possible to walk or bike due to weather, distances, or safety. Many areas do not have a decent public transit system setup and is not provide enough coverage to meet everyone's needs. I am sure many of you probably don't like you co-workers enough to want to sit in a car with them for the commute to and from work, or live in a convenient place to be picked up or dropped off by them; but if you are lucky enough to be in a position where you can walk, bike, take transit, or like your co-workers enough to carpool, you really should.

Buying locally produced goods will also play a major role in reducing the amount of GHGs produced from the transportation sector. Many items, including food, clothing, furniture, technology etc., are made in other countries or even other areas of your country and require boats, airplanes, trains, and trucks to get to the store where you shop. If you buy your food and other items that were produced locally, fewer GHGs will be created to get the products to your stores.

The type of vehicle you drive should depend on your lifestyle. I currently drive a 2013 Toyota Tacoma, double cab (TRD). I drive a lot for work, I spend a lot of time alone in remote areas, sometimes out of cell reception. The Tacoma (similar to a Helix) is a compact truck, powered by gasoline, and is smaller, lighter, and more fuel efficient than many of the larger full-sized trucks; it is also very reliable. I don't think I have done anything to it besides general maintenance (oil changes, brakes, tires, etc.) since I bought it in 2015, and I drive it hard, in every condition imaginable. This suits my lifestyle because there are not enough charging stations currently in the areas where I work, I needed a truck bed to transport equipment, supplies, and gear, and I needed something that is reliable, as I spend a lot of time out of cell phone reception and most of the time by myself. I am excited for the new hybrid trucks to come out. Companies, including Toyota, Ford, Dodge (RAM), and Jeep, are all going to be a part of it. I am most excited for the hybrid Toyota Tacoma, but I will have to wait a few years for the price to drop. People living and working in colder parts of the world have a need for diesel vehicles because it is too cold for gas or electric vehicles to function reliably. If you live in a region that is not too cold and are just driving short distances around town for work, school, or leisure, you should be using an electric or hybrid vehicle, depending on how the power is produced in your area. You should not be using a vehicle that is meant to be used for getting around an open pit mine up north. I am sure people will think you are much cooler if you do your part to help save the planet and get a reasonable vehicle rather than if you have a truck which people must jump to get in and out of, and leaves a cloud of smoke behind it. One thing I always found funny is that it is people who work in industries and live lifestyles that are the worst for the environment who love to spend their time and money on activities they can do outside, on lakes and in forests, like camping, hunting, fishing, etc.; I can never understand it.

Electric and hybrid vehicles have their place, but I do not feel they are the only solution for the transportation industry's contribution to excess greenhouse gas emissions currently. As mentioned in the power generation section above, 60% of the power generated globally (specifically in the U.S.) is from fossil fuels. The amount of greenhouse gases created to charge an electric vehicle could be more than the amount of GHGs created from driving a gas or diesel-fueled vehicle. The average lifespan of an electric car battery can range from 8-15 years, depending on the climate, how the car was driven, and how it was maintained. I feel electric cars are ideal for people living in areas that uses green energy (hydroelectric, solar, wind, tidal, or nuclear) and use the vehicle lightly for something like commuting to and from work every day, as long as work is not too far. If that is not your lifestyle and you do a bit more driving, occasionally heading out of town, etc., then I feel you should be utilizing a hybrid vehicle. I think that, given how the world is currently setup regarding power generation (and will be for the next decade or so), hybrid vehicles are what people should be upgrading to if they are upgrading for environmental reasons. Again, what is best for you really depends on your lifestyle, where you are located, and where you will be driving to, and a fully electric vehicle might be best for you.

Currently, if everyone had electric vehicles, there would be major issues with power generation, not only because 60% of the power generated is from fossil fuels, but the demand for power to charge the vehicles would be higher than the amount of power we can currently produce. I have not found any research on this, and maybe it is something for a PhD student to write their thesis on, but my best guess for an optimal balance of electric vehicles, hybrid and fossil fuel powered cars for the environment and society to live in a balance which meets everyone's needs would look something like 60% electric, 30% hybrid, 10% fossil fuel powered.

There are alternative fuels like hydrogen gas and bio-fuels, which are currently being explored and developed to replace gasoline and diesel. Hydrogen would drastically decrease the amount of GHGs produced from transportation, as water is the only byproduct of using Hydrogen. There is an array of benefits from using different types of bio-fuels, but currently these alternative fuels are not readily available as technologies are still being developed.

Simple changes like letting office workers work from home instead of having to drive to the office would make a huge impact on the amount of GHGs we produce, as well as saving money on fuel and office space, since you will not have to pay for the office space. However, some duties require travel to conduct site visits for an assessment or to provide a quote, to meet a potential client, or to check in on a project. I understand the importance of a handshake and in-person meetings as it shows that you are putting the effort into seeing the people, which helps to strengthen a relationship. If we could somehow change that mentality and get the business mindset to shift away from the need for in-person meetings and to be comfortable with virtual meetings, it would greatly reduce the amount of business travel and the amount of GHGs we produce.

Aviation

It is estimated that approximately 12% of U.S. air travel is for business, and this number has decreased since the pandemic. It looks like most air travel is done for personal use, and recently, many owners of private jets have been causing quite a stir. I get it, if I was craving gelato or calamari, I would much rather go to Italy in comfort, take a little nap on the way, pop some bubbly, and not be surrounded by crying kids, barking dogs, and the smell of other people's socks, but we are in a time where we have to put some of our personal comforts aside for the greater good. I know at least my generation has been very lucky as we have not had to sacrifice as much as previous generations, and we have probably had it the easiest with all the advancements in technology and society, and it is looking like it is time for us to put our time in.

Regarding people who use private jets, there is the matter of security. For example, it would cause quite a commotion if someone like Taylor Swift were sitting in economy class; it wouldn't matter if the airplane and airport were full of "Swifties" or not, I am sure it would cause a safety issue for her. For situations like the celebrities and high-profile businesspeople who require security, it would be beneficial if someone could start a ride share app which would enable these people to essentially airplane pooling (flying version of carpooling) and do their best to reduce the amount of airplanes in the sky and fuel being consumed. They could still fly in their luxury jets, but just enable other people of the same level to catch a ride. Seeing as networking is a key to success in many industries, this would not only reduce the number of planes in the sky and GHGs produced, but also help develop better business relations. Something like this would require an initial large payment for a screening process, and then additional costs per flight to help cover the cost of the flight (fuel, pilots, cabin crew, airport fees, etc.). There could be different tiers, and potentially the options of selecting either a quiet, relaxing environment for rest or work, or a bottle-popping party all the way to your destination.

The train is my favorite way to travel long distances, even though it takes longer to get to your destination when compared to a airplane, you are able to get up and walk around, you can look out the window and enjoy the scenery, you don't have to worry about traffic, and in many cases there is a restaurant / bar car to go and get decent refreshments when required. It is estimated that taking the train instead of flying cuts your CO2 emissions by 90%. In addition to those perks, you can get yourself a bed/room on the train for the longer trips; there is usually some form of entertainment like musicians as well, and the meals can be pretty good. I have met some great people on my train rides around the world, but I also had a few horrible experiences. The worst was when I was in Indonesia on a train which was just meant for locals to travel short distances, not for people to travel across the entire island of Java. I was on the train for about 18 hours with no AC, on a solid wood bench crammed into a tight space in the blistering hot Sun, probably just to save $20; #backpackinglife, #mybad.

Trains are the best choice for transporting consumer goods, everything from grain, clothes, technology, food, vehicles, etc., as they can take a lot more of a load at a time than trucks, and their locomotives produce a lot less GHGs than trucks. There are also advancements in rail transportation, which include full electric locomotives and alternative fuel usage. Rail is by far the most environmentally sound way to transport goods. The only negative is that they are limited to where the tracks go.

I was watching a movie last night called, "The Irishman", released in 2019, with a screenplay by Steven Zaillian, directed by Martin Scorsese, and has a few of your favorite actors like De Niro, Pesci, and Pacino. There is a scene in it where Pacino, who is playing Jimmy Hoffa (an American labour board leader and president of the Brotherhood of Teamsters for 14 years), is speaking to a room full of truck drivers and proclaims, "If you got it, a truck brought it to you". For the most part, that is correct; everything you have bought or were given was most likely transported by a truck to the store where the item was purchased. That truck probably picked up the load from a wholesaler, producer, or a terminal for trains or cargo ships and brought it to the retail store for you to buy. Look around right now and try to think of the voyage your furniture, technology, and clothes went through to get to be in wherever you are. Did it come from a local producer down the street, and you basically just picked it up, walked it over and put it where it is, or was it imported from overseas, sat in a shipping container for months, unloaded at a port, loaded onto a train or truck, transported to the store where you purchased it, and then either had it delivered, picked it up yourself or had to borrow your buddy's truck. As mentioned previously, only around 10.6% of the greenhouse gas emissions from the transportation industry are from international shipping, but there is still room for improvement. The obvious benefit of buying things that are made overseas is that they can be much less expensive than buying something that was made locally. We will talk about the waste management aspect of this later in the book, but generally, the less expensive option can also be the worst quality and will end up in a landfill in no time.

One thing that saves me a lot of money and helps reduce the environmental impact from my shopping is buying used items as much as possible. This includes everything, including furniture, kitchen appliances, tools, clothes, technology, etc. By buying used items, you could reduce the demand on stores to import more items from overseas, or wherever they originate, and indirectly reduce the amount of greenhouse gas emissions it would take to produce and transport the item. Something that saves you money and is good for the environment seems like a win-win to me. We should really get comfortable with reusing, refurbishing, and repurposing items, not only to help create a more circular economy, but also to reduce the amount of items being produced and transported from wherever they are produced. I know many people still believe that they need to have the newest and shiniest thing, and that buying something that is used is something to look down on, but we need to get past that and realize that the benefits of reusing and repurposing items should be our priority. Buying and fixing things up can also be a fun side project or even a hobby. This would not only help save the environment, but save you money; doesn't that seem like a good idea?

Alternate Fuels

Major research and development regarding alternative fuels for vehicles is underway. Per the Environmental and Energy Study Institute, "alternative fuels to gasoline and diesel which provide lower, or zero greenhouse gas emissions include biofuels, Hydrogen, Natural Gas, and Advanced Diesel, as well as electricity". Their descriptions are below regarding alternative fuels.

Biofuels

Several transportation fuels can be produced from biomass, helping to alleviate demand for petroleum products and improve the greenhouse gas emissions profile of the transportation sector. Ethanol from corn and sugarcane, and biodiesel from soy, rapeseed, and oil palm, dominate the current market for biofuels, but several companies are moving forward aggressively to develop and market several advanced second-generation biofuels made from non-food feedstocks, such as municipal waste, algae, perennial grasses, and wood chips. These fuels include cellulosic ethanol, bio-butanol, methanol, and a number of synthetic gasoline/diesel equivalents. The greenhouse gas and oil reduction potential of different biofuels can vary greatly depending on the specific process and feedstocks used. Biofuels can be used in blends with gasoline or diesel or can be used as standalone, drop-in fuels. Until we are able to produce a significant amount of electric vehicles that run on renewably-produced electricity, biofuels remain the only available widespread source of clean, renewable transportation energy.

Hydrogen

A hydrogen fuel cell vehicle is a type of electric vehicle that uses a fuel cell instead of a battery to power the car. Hydrogen fuel cells can be a clean, reliable, quiet, and efficient source of high-quality electric power. They use Hydrogen as a fuel to drive an electrochemical process that produces electricity, with water and heat as the only byproducts. The two main applications for hydrogen fuel cells are in stationary power sources and hydrogen fuel cell vehicles (FCVs).

Hydrogen can be extracted from water using electrolysis, using power from renewable solar or wind, nuclear energy, or fossil energy. It can be extracted from renewable biomass or coal using high-temperature gasification. Using chemical catalysts, it can also be derived from renewable biogas, renewable ethanol or methanol, or fossil natural gas. The environmental impact of hydrogen production depends on the sources and processes through which the Hydrogen is derived. Today, most Hydrogen is derived from fossil natural gas, emitting fossil carbon dioxide as a byproduct.

Fuel cell Vehicles (FCVs) have a driving range like a vehicle with an internal combustion engine – 250 to 400 miles per tank of fuel. FCVs are commercially available for lease, but a hydrogen refueling infrastructure does not exist yet in most parts of the country. California has a few stations in operation, and more under construction. The higher cost of Hydrogen on an energy equivalent basis is outweighed by the greatly increased efficiency of the electric drive system relative to the internal combustion engine.

Natural Gas

Natural gas, when used as a transportation fuel, can emit up to 25 percent less carbon per unit energy than conventional gasoline and can be sourced domestically. Natural gas is currently used in many transit buses, short-haul commercial trucks, and some other vehicle fleets. Use in passenger vehicles is hindered by the added space needed for fuel storage and the limited availability of retail fueling stations, but has been explored by at least one major automaker.

Advanced Diesel

A cleaner generation of diesel fuels and advanced diesel engines together are achieving emission levels that approach those of low-emission gasoline-powered engines, while diesel-powered vehicles can get more than 25 percent better fuel economy than comparable gasoline models. In the United States, diesel fuel is primarily used for trucks, buses, and trains. Half of all passenger vehicles in Europe, however, run on diesel, and clean diesel technology may lead to expanded use in U.S. cars and light trucks.

Companies and municipalities with fleets of vehicles are starting to convert to alternative fuels such as natural gas, and it is possible for individuals to convert their gas and diesel vehicles as well so they can use these alternative fuels. The natural gas people use to heat their homes is not the same as the natural gas used to power vehicles. It would be great if we could figure out a way to use the same gas to heat our homes as well as to power our vehicles to enable people to fill up their vehicles at home, similar to EVs and just pay it off when the gas bill, but the heaters in the home and a vehicles engine have different requirements and currently need different fuels.

An issue regarding these alternative fuel sources is that there are not enough sources available to meet the demand if everyone suddenly switched over. Natural gas is currently the best option out of the alternative fuels, but with more research and development, Hydrogen would be by far the best solution since water vapour is the only byproduct of using it. Natural gas is a better fuel source than gasoline, diesel, or coal, but it has its own issues. Natural gas used to power vehicles is mainly methane, a very potent greenhouse gas, and the extraction, processing, and transportation of this natural gas can cause accidental leaks into the atmosphere and cause other environmental issues. According to the U.S. Energy Information Administration, the two most common ways Hydrogen is produced are through steam methane reforming and electrolysis. Hydrogen fuel is created in large facilities, and depending on which method is used, it can require extracting natural gas from the Earth, specifically methane.

Steam methane reforming separates the Hydrogen from the carbon in methane (CH4) by using high temperatures and pressure as well as a catalyst to help create Hydrogen (H2), Carbon Monoxide (CO), and a small amount of Carbon Dioxide (CO2). The second way hydrogen gas is created is through electrolysis, and this is the way you might have seen in high school. This process uses an electric current to separate the molecules in water (H20) into Hydrogen (H2) and Oxygen (O). As you can see, there are no harmful byproducts from this process; it does require a lot of energy, but if clean energy is used for this, it should not be an issue.

When natural gas is used to create power, it produces CO2 (less than gasoline or diesel), while when hydrogen gas is used to create power, it produces water vapor. Issues with Hydrogen gas include the energy consumption required to create it, it generates less energy when combusted per pound than methane, so it requires more, and it is also much lighter, which makes it more difficult to manage. With more research and development on how to make Hydrogen gas easier to use, it could provide an excellent source of fuel that produces zero greenhouse gas emissions.

I just read an article regarding Helium3 from the European Space Agency, and how it could be the answer to the fuel requirements, but it can currently only be found on the ocean floor or on the moon; mining on the moon is estimated to start in 2030 to try and retrieve the material.

Agriculture / Aquaculture

This book will discuss agriculture as the process of growing crops or raising animals on land, and aquaculture, which includes the raising and breeding of fish, shellfish, and aquatic plants. Both Agriculture and Aquaculture are fundamental aspects of humanity as they produce the food we need to survive, but the methods currently used, mainly for Agriculture are contributing to the greenhouse gas overload and are the fourth highest contributor globally.Aquaculture'Aquaculture's contribution of GHGs in minimal as mentioned in a 2020 study titled, "Quantifying greenhouse gas emissions from global aquaculture", completed by Michael J. Macleod, Mohammad R. Hasan, David H.F. Robb & Mohammad Mamun-UR-Rashid, only 0.5% of the greenhouse gas emissions from human activity in 2017 was caused by aquaculture, while agriculture was responsible for approximately 20%. One thing to keep in mind is that in 2017, approximately 300 million tons of products from aquaculture (including fish, animals, and plants) were produced, while approximately 1.6 billion tons of meat from pigs, chickens, and cattle were produced. Cattle was the highest at approximately 680 million tons of meat produced, then meat chickens at approximately 550 million tons, and finally pigs at approximately 400 million tons. Seeing as the number of products from aquaculture was approximately 19% of the production from agriculture and the greenhouse gas emissions of aquaculture were far less than that, it shows how aquaculture is a much more environmentally friendly food source regarding greenhouse gas emissions over agriculture if done properly.

According to Living Oceans, the following issues come with aquaculture:

Use of Marine Resources

Most of the fish that we love to eat are carnivores. They require great amounts of wild fish as feed to rapidly grow to a large enough size for human consumption. Unfortunately, some of the most popular farmed species require anywhere from 3-15 kilograms of wild fish to grow just 1 kilogram. There are some species, such as scallops and tilapia, that use little to no wild fish in their feed. There is also ongoing research to generate alternative diets that include plant-based materials.

Pollution and Disease

Open net-cage aquaculture provides no barrier between the farmed fish and their surrounding environment. This means that enormous amounts of fish feed, feces, and chemicals are released into the environment every day, often causing inhabitable conditions for other species. In British Columbia alone, salmon farms produce the same amount of waste as a city of half a million people. Fish farms can also create the perfect environment for disease transmission. From open net-cages, these diseases can easily spread to wild fish, wreaking havoc on local populations.

Escapes

Farmed fish that escape into the wild pose a serious risk to wild species. They can spread disease, create unnatural competition, and introduce alien species into an ecosystem. There is also the possibility of interbreeding, forever altering the genetic pool of a population. The number of escapes and their ecological ramifications can be hard to track. However, there are several documented cases of thousands of fish escaping at once.

Habitat Damage

Fish farms are often situated in or near sensitive coastal ecosystems, including estuaries, mangrove forests, and coastal reefs. These ecosystems often provide a nursery for many different species of juvenile marine life, which are particularly susceptible to environmental stresses. These ecosystems can be destroyed to make way for fish farms, or irreparably damaged by the pollution and diseases they can potentially generate.

Management

The aquaculture industry is experiencing a period of rapid expansion. It is important that government and regulatory bodies ensure that this industry is maintaining and enforcing environmental standards. This includes protecting habitat, mitigating pollution, and decreasing the use of wild fish feed. One of the best ways to regulate the fish farming industry is through independent certification. There are currently several different aquaculture certification systems in place, but a universally credible, transparent certification system does not yet exist.

Due to overfishing and the changing aquatic environment killing wild fish, aquaculture (farmed fish) will have to be our main source of seafood until we make the ocean more hospitable and control the amount of commercial fishing better.

Agriculture Globally

Agriculture is unavoidable as we need food to live, but many changes in farming methods due to the need to produce large amounts of crops and livestock to keep up with the growing demand for food have greatly altered the effects of food production. Issues related to the changes in agriculture include an increased use of fresh water for irrigation (approximately 70% of human use), increased production and use of chemical fertilizers, insecticides, and herbicides, destruction of natural lands, decreasing biodiversity, as well as an increase in greenhouse gas emissions from using certain fertilizers heavy in nitrogen. Farmers are also turning to genetically modified (GM) seeds to have more efficient crops, which makes sense, but some studies have found that food grown from some of the GM seeds has some health risks.

Regarding genetically modified (GM) crops, an article from the Journal of Food Science and Technology, written by A.S. Bawa and K.R. Anilakumar, titled "Genetically Modified Foods: Safety, Risks and Public Concerns – a review", states the following:

GM foods are useful in controlling the occurrence of certain diseases. By modifying the DNA system of these foods, the properties causing allergies are eliminated successfully. These foods grow faster than the foods that are grown traditionally. Probably because of this, the increased productivity provides the population with more food. Moreover, these foods are a boon in places that experience frequent droughts or where the soil is incompetent for agriculture. At times, genetically engineered food crops can be grown in places with unfavorable climatic conditions too. A normal crop can grow only in a specific season or under some favorable climatic conditions. Though the seeds for such foods are quite expensive, their cost of production is reported to be less than that of the traditional crops due to the natural resistance towards pests and insects. This reduces the necessity of exposing GM crops to harmful pesticides and insecticides, making these foods free from chemicals and environment-friendly as well. Genetically engineered foods are reported to be high in nutrients and contain more minerals and vitamins than those found in traditionally grown foods. Other than this, these foods are known to taste better. Another reason for people opting for genetically engineered foods is that they have an increased shelf life, and hence there is less fear of foods getting spoiled quickly.

The biggest threat caused by GM foods is that they can have harmful effects on the human body. It is believed that consumption of these genetically engineered foods can cause the development of diseases that are immune to antibiotics. Besides, as these foods are new inventions, not much is known about their long-term effects on human beings. As the health effects are unknown, many people prefer to stay away from these foods. Manufacturers do not mention on the label that foods are developed by genetic manipulation because they think that this would affect their business, which is not a good practice. Many religious and cultural communities are against such foods because they see it as an unnatural way of producing foods. Many people are also not comfortable with the idea of transferring animal genes into plants and vice versa. Also, this cross-pollination method can cause damage to other organisms that thrive in the environment. Experts are also of the opinion that with the increase of such foods, developing countries would start depending more on industrial countries because it is likely that the food production would be controlled by them in the time to come.

The article goes into further details on specific crops like fruits and vegetables, vegetable oils, corn, soy beans, and sugar, as well as other areas regarding GM food, but a key takeaway from the article is as follows:

The GM foods have the potential to solve many of the world's hunger and malnutrition problems, and to help protect and preserve the environment by increasing yield and reducing reliance upon synthetic pesticides and herbicides. Challenges ahead lie in many areas, viz. safety testing, regulation, policies, and food labelling. Many people feel that genetic engineering is the inevitable wave of the future and that we cannot afford to ignore a technology that has such enormous potential benefits.

Issues with decreasing soil fertility due to the hotter temperatures removing moisture from the soils, increased CO2 being stored in the soils, increased run off removing soil from the changes in weather are all hindering farmers from naturally growing high-yielding crops, which makes them rely more on fertilizers and GM seeds. As crops grow, they also remove nutrients from the soil, and if those nutrients are not replaced, the next crop grown in the area will not have the proper food to grow. The United States Department of Agriculture proposes that cover crops be used between the cash crops to help with the following areas:

Decomposing cover crops help feed soil life and contribute to stable soil organic matter, which improves nutrient cycling and soil structure.

Cover crop organic matter can help join soil particles into stable aggregates. These aggregates and cover crop root channels help soils better absorb intense rain, resist erosion, and improve water holding capacity in drier conditions.

Plant cover helps intercept and lessen the force of raindrops. This reduces soil erosion, compaction, and nutrient runoff from intense rainstorms.

Cover crops reduce water pollution risks and remove CO_2 from the atmosphere. They help the environment and reduce a farm's carbon footprint.

Cover crops provide economic benefits. They can increase crop yields, reduce machinery costs, out-compete weeds, break disease and insect cycles, host beneficial organisms, attract pollinators, scavenge nutrients, and supply forage.

Cover crops take up excess water, which can help dry out wet fields before planting.

They also provide the following warnings regarding cover crops:

Seed, fuel, and planting costs may not offset the economic benefits of cover crops in the short-term, but greater benefits are often experienced in the long term.

Cover crops may immobilize nitrogen or deplete soil moisture, causing yield loss in the subsequent crop.

Climate, management, and genetics affect the degree and duration of benefits from cover crops.

Some cover crops may be difficult to terminate, go to seed, and become a weeds if not properly managed.

In 2019, China, India, Brazil, Russia, and the U.S. were the top producers of agricultural products, with the U.S. being the largest exporter. In 2019, China was the largest exporter of wheat, rice, and many vegetables, including onions and cabbage, and second corn internationally. Below are the highlights from an article called "Agricultural Production Statistics 2000-2021" by the Food and Agriculture Organization of the United Nations:

The global production of primary crop commodities reached 9.5 billion tons in 2021, increasing by 54 percent since 2000 and 2 percent since 2020.

The global production of cereals went up 64 million tons, or 2.1 percent, between 2020 and 2021, driven by a 4.1 percent increase in maize production. Maize, wheat and rice accounted for 90 percent of the total cereal production in 2021.

The world production of sugar crops increased marginally between 2020 and 2021. Sugar cane is the main sugar crop, with 1.9 billion tons in 2021, compared to 270 million tons in 2021 for sugar beet.

Roots and tubers global production rose by 1.9 percent between 2020 and 2021, mostly due to an increase in cassava and potato production.

In 2021, world fruit and vegetable production reached 910 million tons (+1.1 percent from 2020) and 1.2 billion tons (+1.4 percent), respectively.

The production of oil palm fruit, soya beans, and rapeseed, the main oil crops contributing to vegetable oils, reached a volume of 859 million tons in 2021, increasing by 2 percent from 2020.

Chicken, pig, and cattle were the main meats produced worldwide, accounting for 316 million tons in 2021.

China's pig meat production recovered in 2021 after the low volumes of 2019–2020 caused by the African swine fever outbreak.

Due to the need for farmers to produce more food in the same area, they need to rely on modified seeds and fertilizers to make their crops more efficient. A major issue with the fertilizers is that they can contain nitrogen since nitrogen is essential for plant growth as it helps with the production of amino acids, nucleic acids, and proteins. Unfortunately, when there is excess nitrogen in the ground, it can undergo a chemical reaction with air and become Nitrous Oxide, which, as we know, is a potent greenhouse gas. Different countries produce different agricultural products like beef, pigs, wheat or corn and different farms within the countries might produce different products as well. This means that not only each country, but each farm must look at what they are doing and what changes they must make to be more sustainable. Governments could create policies which control the types of fertilizer allowed or try to control the number of livestock in their country, but the responsibility of how food is produced should fall onto the farmer, and the consumer (you and me) need to choose to buy the more environmentally friendly products to help persuade farmers to change their processes. The European Environment Agency has an interesting article titled, "Soil, land and Climate Change" last modified on August 29, 2023, which discusses the importance of soil health, how it is a major carbon sink, and the importance of maintaining healthy soils to help combat climate change.

The other major contributor of GHGs from the agriculture sector is from the methane created in the livestock's digestive system and released. Science and technology can play a major role in reducing the amount of methane produced by livestock through altering the feed of the animals or promoting nutrients that aid in gut health. Changing the demand on the types of food we eat will also play a major role; this would include reducing the amount of beef you eat and replacing it with other sources of animal protein which cause less GHGs like eggs, chicken, turkeys or something like goat, kangaroo to help bring the demand of beef down as it would bring the production of beef down; fun fact, kangaroos stomach has bacteria which inhibits the production of methane. You can also remove animal protein from your diet completely and rely on plant-based foods, which are high in protein like lentils, chickpeas, quinoa, tofu etc. to reduce the demand on animal farming. I will just try to eat less beef and more eggs, chicken, turkey, and goat, but if you are able to completely switch to a vegetarian diet, good for you! If you identify as a carnivore or like a diet heavy in meat, it would be very beneficial if you committed fully to that world and included all the animal parts like the liver, kidneys, heart, tongue, and not just the juicy stakes. The liver, kidneys and heart are all great sources of nutrients like iron, and even the bone marrow is delicious, a great source of collagen and healthy fats. Bone marrow is inexpensive when you buy the bones at the store with the marrow inside and super easy to make (just heat up the bone on the BBQ or oven, then use a spoon to remove it when it's warm and soft, eat it as is or spread it on toast. By using more of the animal, we are reducing the amount of waste the producers create and helping to improve the environment.

On the flip side, there are also negative environmental impacts of increasing production of plant-based foods; these impacts are from deforestation and draining of wetlands for more agricultural lands and the additional use of nitrogen-containing fertilizers. From a greenhouse gas perspective, changing your diet to a plant-based diet or reducing the demand on conventional meat products would make a significant positive difference. I am trying to eat more plant-based foods, which are high in protein; my go-tos are lentils, quinoa, and chickpeas. You must research good recipes that include these, and my go to is use them in chili as a meat substitute; I am also eating more eggs, chicken, turkey, and goat while cutting back on my beef. I don't think I could ever completely remove animal meat from my diet. I tried it once for a few months, and it really didn't work for my body type. Seeing as beef prices are increasing, modifying your diet to be more environmentally sustainable will also save you money. Hunting and fishing for your own protein can also be very beneficial for the environment as long as it is done properly and not killing too many. In many areas of the world, certain animal populations have increased to harmful numbers as they are over-consuming vegetation and even getting into the food being grown for human consumption. Hunting can help with population control, reduce the demand on the production of beef, and save families money.

Solving the issues regarding agriculture is tricky as a farm setup for cattle one day can't just become a soy bean farm the next; livestock farms are setup a certain way to produce and take care of specific types of animals, and crop growing farms have a specific environment like nutrient-rich soils to grow crops. A farm that is setup to raise cattle can't just start raising chickens without spending a lot of money constructing new buildings, bringing in new technology, and learning an entirely different skillset. I would be just as difficult for a livestock farm to switch over to growing crops. There are a lot of things that must fall into place to have successful crops, and just because a farm has a lot of land doesn't necessarily mean that it can grow crops. The last thing we would want to do is to put farmers out of business, they have been and will be key players in the survival of humanity, and we have a lot to be grateful to them for.

The main way to slow down the production of beef is by altering how we choose to shop and eat; what we create a demand for is what will be produced. One way to help reduce the environmental impacts from beef farming is to figure out where the beef is coming from and how it was produced, and then choose to only purchase beef from a sustainable source. If it were produced in somewhere like Alberta, Canada, it would most likely have caused less of an environmental impact than if it were from Brazil. If we keep on buying and consuming beef, farms will continue to produce beef, but if we change our demand to another source of animal protein (eggs, chicken, turkey, fish, goat etc.), the production of that animal will increase and the production of cows will decrease. We still need to make sure we are keeping farmers employed, and we must figure out ways to help them slightly alter how they operate to ensure they are still generating income and producing enough food for the growing population. As mentioned before, it is difficult and expensive for a farm to change directions and start producing something completely different from what they are setup to do, but it wouldn't be so bad if they just slightly shifted their course to something similar, like changing from cows to goats. A publication by Nazan Koluman, titled, "Goats and their role in Climate Change", from the 26th of September, 2023, as well as a publication by Pratap Pragna, Surinder S Chauhad, Veerasamy Sejia, Brian J. Leury, and Frank R. Dunshea, titled, "Climate Change and Goat Production: Enteric Methane Emission and its Mitigation", both go into detail on the benefits of goats as livestock. Koluman found that

Because goats have a high feed conversion efficiency, they emit less enteric methane per unit of feed consumed. Genetic selection of goats with higher feed conversion efficiency generates a reduced amount of CH4. Further, genetic selection for less CH4 production indirectly helps the farmer increase their profits without any extra carbon credits by increasing the feed.

Pragna et al state the following:

Small ruminant, and in particular goat, farming is very important because of the relatively low input requirements and the corresponding high expected output. Furthermore, goats emit less enteric methane (CH_4) than all other domestic ruminant animals per unit body weight.

The amount of methane any animal creates has a lot to do with what they eat, their genetics, and their gut health. New Zealand has achieved some amazing advancements with breeding their sheep to produce less methane, and as they are a large producer of sheep meat globally, that is a major win. The reduction in methane production might be offset by the transportation of the products to their global buyers, so it would be ideal if we could get a local population of their low methane-producing livestock. There are many benefits for farmers to switch over to goats. The main ones are that they are less expensive to maintain, they require less water and food, they do less damage to the grazing areas, and they are better at adapting to the hotter temperatures and climates we are getting into. As previously mentioned, another great benefit for consumers of switching from beef to other animal products is cost savings. In many places, it is much cheaper to buy eggs, chicken, turkey, and goat than it is to buy beef products. I have seen entire frozen goats for sale for under a couple hundred dollars; that is expensive initially, but if you find a local butcher in your area who will be able to cut it up into the required portions, you can refreeze it and have enough meat to last you a very long time, or split the cost and the meat with friends and family saving you money in the long run. Frozen meat can be thawed and refrozen in a way that is safe to eat as long as it is done properly; it is also possible to get sick from doing this, but your butcher should know how to do it properly.

Alternative sources of protein to farmed animals, which are currently being researched and developed, are plant-based meat substitutes (plant-based meat analogues), insects, and cultured meat. With the growing demand for food due to the increasing population, the decreasing amount of available land, water, and resources, as well as the growing interest in animal welfare, these alternative options are looking like they will be in higher demand. An article by Hyun Jung Lee, Hae In Yong, Minsu Kim, Yun-Sang Choi, and Cheorun Jo, titled, "Status of meat alternatives and their potential role in the future meat market- A Review", copywritten in 2020 by Asian-Australasian Journal of Animal Sciences, explores the need and benefits of these alternate meat sources. For clarity, the table below from their paper provides definitions of the alternate meat sources being discussed:

Types	Definition
Conventional meat	Traditional meat from farm animals
Plant-based meat analogue	Meat analogue made of plant and fungus proteins.
Edible insect	Insects used as food resources
Cultured meat	Artificial meat produced using stem cell technology
Modified meat	Meat from genetically modified animals
3D-printed meat	Fabricated meat made of native or non-native food materials with a 3D printing system

The publication mentions the benefits of plant-based meat substitutes below:

In the current market, several products are successful as plant-based meat analogues and seem to provide sufficient amounts of proteins to our diet as meat alternatives. Bohrer investigated the nutritional contents in four major types (beef burger products, beef meatballs, pork ham, and chicken nuggets) of traditional meat and plant-based meat analogues in market. They found that each beef patty in a burger contains 23.33 g of protein, whereas a meat analogue patty has approximately 19.46 g of protein. However, plant-based meat analogues have less cholesterol and more dietary fiber, which can be appealing to consumers. The other types of products (beef meatballs, pork ham, and chicken nuggets) also showed similar overall results. Therefore, as far as nutritional aspects are concerned, especially the protein content, plant-based meat analogues are likely to be good substitutes for traditional meat.

I have not been able to make the switch over to plant-based meat substitutes yet, mainly because I have the memory of the old cardboard-tasting and textured veggie burgers, which were either too mushy when undercooked or made me feel like I was eating a hockey puck when it was overcooked, and the taste was very bland. Luckily, this industry has completed a lot of research and development, and a few producers have seemed to perfect the production of the plant-based meat substitutes, especially with common items like burgers and nuggets, matching the taste and texture of the beloved animal meat products. If you are interested in helping to reduce the demand for livestock production, you should give it a try. Some are better than others, so just do some research before you try it, or try different types to see if you like one more than the other.

Food made from insects is another top alternative to farmed meat; for many of us, it is not a normal thing to think of as food, but it has been a staple in many culture's diets around the world throughout history. In the publication by Jung Lee et al., they describe the following benefits to using insects as an alternative option to meat:

The major purpose of the consumption of insects by humans is to provide an excellent source of proteins. The nutritional values of edible insects vary depending on their species, sex, metamorphosis state (e.g., larvae, pupae, and adults), origin, diet, and different methods of processing due to their large diversities. Xiaoming et al reported that protein content in 100 different species ranged from 13% to 77% on the basis of their dry matter. Analyses of 87 insect species in Mexico revealed a protein content of 15% to 81% with high digestibility reviewed the nutritional value of frequently consumed insects (e.g., beetles, files, bugs, bees, wasps, sawflies and ants, butterflies and moths, grasshoppers, crickets, and locusts) and found large variations (1% to 81%) among the protein contents. The bioavailability of insect protein is also high with good digestibility (76% to 96%), which is a little less than that of egg or beef protein (95% and 98%, respectively) . Thus, undoubtedly, insects can serve as a fine protein source in our diet. In Central Africa, there was a time when about 50% of dietary proteins were obtained from insects. Compared to plant protein, insect protein has nutritional benefits with respect to total protein levels, essential amino acids, and bioavailability. Kouřimská and Adámková stated that some species of insects have high lysine, tryptophan, and threonine contents, which are not found in some plants.

Using insects for food might seem obscene to you, but as mentioned, many people around the world have been using insects as a major source of protein for thousands of years. Below is what

M. Shahbandeh discusses in their publication from March 18, 2024, titled, “Edible Insects – Statistics & Facts”:

Millions of people from around the world have relied on insects as an essential source of nourishment for thousands of years. Insects are high in protein, easy to cultivate, have a low carbon footprint, and could be instrumental in alleviating world hunger and climate change. The Food and Agriculture Organization of the United Nations has endorsed edible insects and encouraged their adoption in the diets of people all around the world. One of the main benefits of cultivating edible insects is that they require fewer resources than traditional protein sources for the same quantity of protein. While producing one gram of protein from beef requires 254 square meters of farmland, producing a gram of protein from insects requires only 18 square meters.

Jung Lee et al also put together the following table to help show the amounts of protein per insect:

Nutrient composition of edible insects depending on different species (on a dry matter basis)

Edible insects	**Protein (%)**	**Fat (%)**	**Fiber (%)**	**NFE (%)**	**Ash (%)**	**Energy content (Kcal/100 g)**
Blattodea (cockroaches)	57.30	29.90	5.31	4.53	2.94	-
Coleoptera (beetles, grubs)	40.69	33.40	10.74	13.20	5.07	490.30
Diptera (flies)	49.48	22.75	13.56	6.01	10.31	409.78
Hemiptera (true bugs)	48.33	30.26	12.40	6.08	5.03	478.99

Edible insects	Protein (%)	Fat (%)	Fiber (%)	NFE (%)	Ash (%)	Energy content (Kcal/100 g)
Hymenoptera (ants, bees)	46.47	25.09	5.71	20.25	3.51	484.45
Isoptera (termites)	35.34	32.74	5.06	22.84	5.88	-
Lepidoptera (butterflies, moths)	45.38	27.66	6.60	18.76	4.51	508.89
Odonata (dragonflies, damselflies)	55.23	19.83	11.79	4.63	8.53	431.33

Edible insects	Protein (%)	Fat (%)	Fiber (%)	NFE (%)	Ash (%)	Energy content (Kcal/100 g)
Orthoptera (crickets, grasshoppers, locusts)	61.32	13.41	9.55	12.98	3.85	426.25

This concept is having issues taking off for the obvious reason that many people do not like the idea of eating insects. It is a good source of nutrition; it is much less expensive and easier to produce than traditional meat products, and it has a significantly lower environmental impact, including greenhouse gas emissions, than other options. I am also hesitant to try a bug burger, but perhaps if they work on the texture and taste of their products, like what they did with the plant-based meat alternatives, I will give it a try.

The third main alternative to farmed meat is cultured meat, which is meat created in labs or other facilities. Per GFC Global, cultured meat is created by taking stem cells from animals like cows, chickens, and ducks, giving them amino acids and carbohydrates, which enables the stem cells to grow and multiply into what looks like ground meat. This form of food is actual meat, the same as you would get from a farmed animal, and consequently, it is not for vegetarians. This is a new method and still being researched and developed, but it is a form of meat production which has a far less environmental impact than conventional meat production. If you refuse to give up animal meat but still want to make a positive impact on the environment, this might be a solution for you.

Minimizing food waste will play a major role in reducing the amount of food we need to produce, reducing the amount of GHGs created from the production of food, reducing the amount of methane produced in landfills from the food waste decomposing, and it will also save you money.

The World Wildlife Foundation proposes the following tips to help you do your part to reduce food waste:

- **Plan ahead and buy only what you need.** Going to the store without a plan or on an empty stomach can lead to buying more than we need. To keep your kitchen on track, try to eat leftovers, think of meals you might eat out, and avoid unnecessary purchases by planning your grocery list ahead of time.
- **Use your freezer.** While there are plenty of benefits to eating fresh food, frozen foods can be just as nutritious. They also stay edible for much longer. A lot of seafood, for example, is frozen before it reaches your supermarket and then thawed and put on display. That means it will only stay fresh for a few days. By buying frozen seafood, you can extend the shelf life of the product considerably. Cooking and freezing food—especially produce—before it goes bad is a great way to avoid having to toss it.
- **Be creative with leftovers.** Before you shop, use the food you already have. Websites like Big Oven, Supercook, and MyFridgeFood allow you to search for recipes based on ingredients already in your kitchen. You can also use

apps like Epicurious and Allrecipes to make the most of what's in your fridge and pantry.

- **Blend, bake, or boil.** Fruits and vegetables that are beyond ripe may not look pretty, but that doesn't mean they can't still taste delicious in recipes. Try using your wilting, browning, or imperfect produce to make sweet smoothies, bread, jams, sauces, or soup stocks.
- **Talk it up.** Preventing food waste is the most effective way to shrink its impact on the planet. If we avoid producing food that we don't eat, we can save the land, water, and energy that would have been used to make it. And awareness is a good first step; according to ReFED, educating consumers about food waste could prevent 7.41 million tons of greenhouse gas emissions.

Industrial

The industrial industry is the final industry of the top four global greenhouse gas producers to discuss before we dive into the major natural carbon sinks. This area includes industrial operations such as manufacturing (things like cement, steel, and chemicals), construction, and mining. China is the highest producer of GHGs in this industry, and that makes sense as they are the main producers of many things we need to function and develop, like steel and cement. Without China producing such large quantities of these products, we would run into many issues regarding construction as the supply would drastically decrease, and prices would inflate, making it much more difficult to build and develop our infrastructure. There is a recurring trend which is apparent in each of the sections we have discussed so far, and that is that the main contributors of GHGs are from industries which are providing us with things we are asking them to make, things that we create a demand for. Good news is that China is forecasting to reach their peak greenhouse gas emissions within the decade (meaning they will start producing less than in previous years) and reach their carbon neutral goal by 2060.

Cement manufacturing is one of the largest producers of CO2 within the industrial sector. The following definition of cement was pulled from Britannica Encyclopedia:

Cement, in general, is an adhesive substance of all kinds, but, in a narrower sense, the binding materials used in building and civil engineering construction. Cements of this kind are finely ground powders that, when mixed with water, set to a hard mass. Setting and hardening result from hydration, which is a chemical combination of the cement compounds with water that yields submicroscopic crystals or a gel-like material with a high surface area. Because of their hydrating properties, constructional cements, which will even set and harden under water, are often called hydraulic cements. The most important of these is Portland cement. Cements may be used alone (i.e., "neat," as grouting materials), but the normal use is in mortar and concrete, in which the cement is mixed with inert material known as aggregate. Mortar is cement mixed with sand or crushed stone that must be less than approximately 5 mm (0.2 inch) in size. Concrete is a mixture of cement, sand or other fine aggregate, and a coarse aggregate that, for most purposes is up to 19 to 25 mm (0.75 to 1 inch) in size, but the coarse aggregate may also be as large as 150 mm (6 inches) when concrete is placed in large masses such as dams. Mortars are used for binding bricks, blocks, and stone in walls or as surface renderings. Concrete is used for a large variety of construction purposes. Mixtures of soil and Portland cement are used as a base for roads. Portland cement is also used in the manufacture of bricks, tiles, shingles, pipes, beams, railroad ties, and various extruded products. The products are prefabricated in factories and supplied ready for installation.

The following description was taken from a publication by Nabilla Mohamad, Khairunisa Muthusamy, Rahimah Embong, Andri Kusbiantoro, and Mohd Hanafi Hashim from January 7, 2022, titled "Environmental Impact of Cement Production and Solutions: A Review."

Precisely to produce Portland cement, raw materials should be combined in the process; an open mining method is used for mining operations. Quarrying is achieved by digging, blasting, and using heavy earth-moving machinery such as bulldozers and dump trucks. A sufficient amount of calcium, silicon, aluminium, and iron parts is needed to form the typical clinker composition. Lime and silica are the primary qualities of cement, while iron decreases the temperature of the reaction and gives the cement its characteristic grey colour. In the cement manufacturing process, limestone, shale, and clay are ready, dried, crushed, blended, and heated in cement ovens up to 1200° C–1450° C to produce the clinker. A high temperature is needed for chemical reactions to occur in the burning region. A nodular substance called a clinker is formed and is then allowed to leave the oven. Clinker is then used as the main ingredient for cement. It is crushed and blended with other ingredients such as limestone and gypsum to produce ordinary Portland cement [18]. During the process, a significant amount of fossil fuels and alternative fuels are burned to provide heat to cement ovens, dryers, and preheaters. In addition, a lot of energy is consumed in cement ovens and crushing machines in particular. Thus, in order to manufacture a large volume of cement, an immense amount of energy is consumed. The initial stages of the cement production process until the final stage contribute to pollution, especially air pollution. Studies have shown cement production releases roughly 0.8 tonnes of CO_2 per tonne of cement to the atmosphere [19]. In the meantime, it is reported that a tonne of Portland cement production produces around a tonne of CO_2 greenhouse gas emissions, and about 2% to 8% of global power consumption is due to the cement manufacturing process.

As you can see, to reduce the environmental impact of cement manufacturing, we need to come up with a substitute for cement or alter the way cement is manufactured.

Below are 6 sustainable substitutes for concrete mentioned in an article by Renewable Energy Magazine, from June 5, 2023, titled "Six Sustainable Alternatives to Traditional Concrete (and how to use them)".

1. Hempcrete

Hemp has become a sustainable alternative in numerous industries. For example, you can find clothes made from hemp in textile factories. Manufacturers also use hemp for paper, food, biofuel, and concrete. You only need 60 to 90 days to produce hemp fibers, and you can grow this crop nearly anywhere in the world. The quick production makes it accessible for industries like construction.

Hemp concrete, or hempcrete, uses hemp hurds and binds them with lime, water, and additives like sand to make the concrete alternative. Hempcrete uses much less water than typical concrete and is much lighter, making it easier for construction crews to handle.

In construction, contractors often use hempcrete for insulation. Lime is dense, giving the hempcrete excellent thermal properties. You can use hempcrete for floors, walls, and roofs to control your house's temperature. Home and office buildings often use hempcrete for partition walls. These walls separate rooms that need different environments, such as a bathroom and a common area.

2. Self-Healing Concrete

Over time, most concrete structures form cracks. The most common causes of cracks are shrinkage and stress. When poured, the concrete may have dried too quickly, or the contractors may have added too much water. Regardless, cracks are the last thing you want to see. How can you repair them? One solution is to use self-healing concrete.

As the name suggests, self-healing concrete repairs itself when cracks occur. This material uses bacterial reactions to mend damage and extend the concrete's life. Self-repair means that moisture and chemicals are less likely to deteriorate the self-healing concrete. You'll spend much less time on maintenance and reduce resource consumption by reducing the need for repairs.

A 2019 Journal of Building Engineering Study examined self-healing concrete, concluding that it benefits construction through sustainability, environmental friendliness, and energy-saving traits. Contractors often use self-healing concrete for bridges and roads because of the frequent heavy loads. You can also see it in marine structures like harbors and piers due to the harsh moisture and salty conditions that accelerate concrete's deterioration.

3. Ashcrete

Another popular concrete alternative is ash concrete or ashcrete. This material combines fly ash from coal combustion with lime and water to produce a durable structure. Ashcrete is sustainable because it repurposes coal waste and reduces the energy-intensive process of typical concrete production. It also uses less water. Research shows that fly ash concrete requires about 10% less water than traditional concrete.

Contractors use fly ash concrete as a partial substitute for Portland cement. Ashcrete is dense and durable, ideal for road construction and structural elements like columns and beams. You can also use ashcrete to produce masonry units like blocks and bricks for walls, partitions, and more.

4. Fiber Cement

Concrete is popular in the building sector because it's low-maintenance. During its lifetime, you won't have to perform much upkeep. When finding concrete alternatives, selecting materials with similar durability is imperative. That's what you get with fiber cement.

This concrete alternative typically contains Portland cement, water, fly ash, and wood pulp. The wood pulp contributes to durability because its cellulose fibers act as reinforcement. Fiber cement resists cracking and warping in most climates because of the wood fibers' strength. For example, fiber cement is optimal in coastal climates because it resists salt, which generally causes erosion.

Fiber cement's durability makes it popular for siding and roofing. Many contractors use fiber cement as exterior cladding in residential and commercial structures. You can also employ fiber cement for interior cladding, providing high-quality finishes for walls and ceilings.

5. Ferrock

Many alternatives aim to reduce the environmental impact of concrete. What if you could eliminate the adverse effects and turn them into positives? You can achieve carbon negativity through

Ferrock. This type of concrete comes from recycled steel and glass to produce a strong material. A 2020 study finds Ferrock has better compressive strength than traditional concrete.

Ferrock is relatively new in construction, so experts are still finding new ways to employ it. So far, Ferrock has been helpful for pavements and walkways to endure high foot traffic. You can also use Ferrock for load-bearing structures like walls, beams, and columns. The compressive strength offers excellent durability in construction.

Ferrock has become popular because of its environmental benefits. The material absorbs CO2 from the atmosphere and converts it into stable minerals. This sequestration process makes Ferrock carbon negative, removing CO2 from the air.

6. Bamboo Concrete

Bamboo has quickly risen as a sustainable material in construction. Contractors use the plant for flooring, roofing, scaffolding, and more. Bamboo is sustainable because it grows fast in nature, and you can regenerate it after harvesting. Once established, bamboo only needs about two or three months to reach maturity. Its sustainability has made it an attractive option for builders looking for concrete alternatives.

Bamboo concrete is popular because it has excellent tensile strength. It's lighter than traditional concrete and has terrific flexibility because of its internal fibers. Using bamboo concrete is wise in Earthquake-prone and humid areas due to its durability and resistance to corrosion. A 2021 Materials Today study finds that bamboo concrete provides strength, durability, and flexibility when used as a reinforcement material.

Bamboo's rise has led numerous contractors to employ this material on building sites. Contractors use bamboo concrete for flooring for slabs or tiles and roofing for shingles or tiles. You can also use bamboo concrete outside for pavements, walkways, garden borders, and other elements.

Recycling concrete and using a type of chemical called polymers to help bind the aggregate are also possible solutions to help reduce the production of cement and the GHGs created from cement manufacturing. An article by Fast Company, from October 18, 2023 titled, "The World has a concrete problem. These quirky alternatives want to be a solution" is an interesting read, but the key take aways are that how some companies are removing CO2 from the air and storing it in their concrete, replacing some of the cement with a type of silica that is created with CO2 capture from their facilities, diverting it from power plants, or using limestone calcined clay cement which could reduce emissions by up to 40% compared to portland cement.

Steel manufacturing is another major contributor to greenhouse gas emissions. An article written by Jinsoo Kim, Benjamin K Sovacool, Morgan Brazilian, Steve Griffiths, Junghwan Lee, Minyoung Yang, and Jordy Lee, from March 8, 2022, titled "Decarbonizing the iron and steel industry: A systematic review of sociotechnical systems, technological innovations and policy options. Says the following about steel manufacturing:

In the manufacturing of these essential goods, iron and steel huge energy inputs are required. The iron and steel sector used 33.57 Exajoules of energy in 2018, and energy cost constitutes a significant portion of steel manufacturing costs, ranging from 20% to 40%, which explains why many decarbonization options are related to energy saving. Critically, the iron and steel industry is the second largest consumer of coal, next to electricity generation. Coking coal is used for chemical reactions in furnaces to make steel from iron ore, so up to 75% of the energy content used in steel production is consumed in the blast furnace. The remaining 25% offers heat at the sinter and coking plants.

Thus, it is perhaps inevitable that the iron and steel industry is highly responsible for global GHG emissions and thus contributes to climate change. The iron and steel sector emits 2.6 Gt CO_2e annually, which is 7% of the global emissions from the energy use and 7–9% of global anthropogenic CO_2 emissions—the highest among heavy industries. Iron and steel are also considered one of the hardest industries to decarbonize due to high heat requirements, using carbon as a process input, low profit margins, high capital intensity, long asset life, and trade challenges. There are no easy ways to create large amounts of heat energy for many iron and steel processes without also releasing CO_2 emissions, and coal is often used both as a source of heat and as part of the production processes. Similarly, the decades-long life cycles of iron and steel plants, the lack of clear financial incentives for decarbonization, and price volatility make it difficult to incorporate carbon-reducing technologies.

World Steel Association states that they are focusing on maximizing the recycling of waste steel and mentioned the following:

To achieve the drastic reductions needed, an entirely new, transformative approach to ironmaking is required, and there are several promising approaches to reduce iron ore at an industrial scale without the release of CO2. These fall into three broad categories: carbon, hydrogen, and electricity.

Carbon Sinks

The Ocean

The largest natural carbon sink is the ocean, and it absorbs more CO2 than all the forests combined. The ocean has a couple of major things working for it. The first is that the CO2 can be absorbed by the water. Secondly, when the CO2 is in the water, a majority of it undergoes a chemical reaction with the water and changes the carbon dioxide into carbonic acid; carbon dioxide (CO2) + water (H2O) -> carbonic acid (H2CO3). The third thing that happens to the CO2 in the ocean is that it is used in the photosynthesis process in aquatic vegetation and aquatic life like phytoplankton, just as trees and vegetation absorb the CO2 on land. Unfortunately, since it has been removing so much CO2 over the past 200 years (since the Industrial Revolution), the ocean is becoming too acidic from all the carbonic acid created from the absorbed CO2, and is becoming an inhospitable environment for some aquatic life; this process is called ocean acidification. There are many serious issues with ocean acidification, including killing off the aquatic vegetation, which is currently using the excess CO2 in their photosynthesis processes, reducing the ocean's ability to remove CO2.

The National Oceanic and Atmospheric Administration (NOAA) of the U.S. Department of Commerce compiled a good article regarding ocean acidification, which mentions the following:

Ocean acidification is already impacting many ocean species, especially organisms like oysters and corals that make hard shells and skeletons by combining calcium and carbonate from seawater. However, as ocean acidification increases, available carbonate ions (CO32-) bond with excess hydrogen, resulting in fewer carbonate ions available for calcifying organisms to build and maintain their shells, skeletons, and other calcium carbonate structures. If the pH gets too low (acidic), shells and skeletons can even begin to dissolve.

Changes in ocean chemistry can affect the behavior of non-calcifying organisms as well. The ability of some fish, like clownfish, to detect predators is decreased in more acidic waters. Studies have shown that decreased pH levels also affect the ability of larvae to locate a suitable habitat. When these organisms are at risk, the entire food web may also be at risk.

While some species will be harmed by ocean acidification, algae and seagrasses may benefit from higher CO_2 conditions in the ocean, as they require CO_2 for photosynthesis just like plants on land. There are some ongoing studies examining whether growing seaweed can help slow ocean acidification.

Estimates of future carbon dioxide levels, based on business-as-usual emission scenarios, indicate that by the end of this century the surface waters of the ocean could have a pH around 7.8 . The last time the ocean pH was this low was during the middle Miocene, 14-17 million years ago. The Earth was several degrees warmer, and a major extinction event was occurring.

Ocean acidification is currently affecting the entire ocean, including coastal estuaries and waterways. Billions of people worldwide rely on food from the ocean as their primary source of protein. Many jobs and economies in the U.S. and around the world depend on the fish and shellfish that live in the ocean.

Current research

Ocean acidification is one aspect of global climate change. Anything we do to mitigate climate change today will benefit the future of the ocean as well. Over the last decade, there has been much focus in the ocean science community on studying the potential impacts of ocean acidification. NOAA's Ocean Acidification Program serves to build relationships between scientists, resource managers, policy makers, and the public in order to research and monitor the effects of changing ocean chemistry on economically and ecologically important ecosystems such as fisheries and coral reefs.

Because sustained efforts to monitor ocean acidification worldwide are only beginning, it is currently impossible to predict exactly how the impacts of ocean acidification will cascade throughout the marine food web and affect the overall structure of marine ecosystems. With the pace of ocean acidification accelerating, scientists, resource managers, and policymakers recognize the urgent need to strengthen the science as a basis for sound decision-making and action.

NASA's Earth Observatory via Holli Riebeek published an article called "The Ocean's Carbon Balance", published on July 1, 2008, and it mentions the following information:

For eons, the world's oceans have been sucking carbon dioxide out of the atmosphere and releasing it again in a steady inhale and exhale. The ocean takes up carbon dioxide through photosynthesis by plant-like organisms (phytoplankton), as well as by simple chemistry: carbon dioxide dissolves in water. It reacts with seawater, creating carbonic acid. Carbonic acid releases hydrogen ions, which combine with carbonate in seawater to form bicarbonate, a form of carbon that doesn't escape the ocean easily.

As we burn fossil fuels and atmospheric carbon dioxide levels go up, the ocean absorbs more carbon dioxide to stay in balance. But this absorption has a price: these reactions lower the water's pH, meaning it's more acidic. And the ocean has its limits. As temperatures rise, carbon dioxide leaks out of the ocean like a glass of root beer going flat on a warm day. Carbonate gets used up and must be re-stocked by upwelling of deeper waters, which are rich in carbonate dissolved from limestone and other rocks.

In the center of the ocean, wind-driven currents bring cool waters and fresh carbonate to the surface. The new water takes up yet more carbon to match the atmosphere, while the old water carries the carbon it has captured into the ocean.

The warmer the surface water becomes, the harder it is for winds to mix the surface layers with the deeper layers. The ocean settles into layers, or stratifies. Without an infusion of fresh carbonate-rich water from below, the surface water saturates with carbon dioxide. The stagnant water also supports fewer phytoplankton, and carbon dioxide uptake from photosynthesis slows. In short, stratification cuts down the amount of carbon the ocean can take up.

In summary, the Ocean has been removing CO2 from the atmosphere and either converting it to carbonic acid, bicarbonate, or aquatic life like phytoplankton, seaweed, and seagrass has been using it in photosynthesis and removing it from the water.

Warmer water absorbs less CO2 than cooler water, and as the ocean's currents slow down, the water becomes less turbulent, mixes less, and prevents the CO2-packed ocean water at the surface of the ocean from getting replaced by cooler ocean water, which is carbonate-rich and contains less CO2 from deeper in the ocean. This process will slow the amount of CO2 the oceans will be able to absorb as the top layer will eventually become too full of CO2 and carbonic acid to absorb the CO2 in the air, and the environment will not be suitable for aquatic life like phytoplankton to survive and the absorption of CO2 for photosynthesis will decrease dramatically, eventually we will lose our biggest carbon sink. The article goes on to bring up some seemingly contradictory findings:

"Ozone naturally warms the upper atmosphere because it captures the radiation from the Sun and re-emits it there," says Le Quéré. "If you deplete ozone, you get a very large cooling in the upper atmosphere." The huge temperature difference between the ozone hole and the rest of the stratosphere causes strong winds around Antarctica. Uneven warming in different parts of the southern hemisphere from recent global warming also created a temperature difference that strengthened the winds. The stronger winds enhance deep water upwelling, which allows carbon to vent into the atmosphere from carbon-rich deep water. In essence, while the ocean may be taking up more anthropogenic carbon to keep pace with levels in the atmosphere, it's also venting more carbon than it did in the past, and that changes the size of the overall sink.

Like Feely saw in the equatorial Pacific, stronger winds made the Southern Ocean vent more carbon dioxide in areas where deep water upwelled to the surface. This idea, that upwelling water releases carbon dioxide, ran counter to what oceanographers had believed about stratification for decades. "When I started, everybody said if the ocean stratifies, then it will absorb less anthropogenic CO_2. But now, it's not so clear," says Le Quéré. If global warming causes upwelling areas like the high latitudes or the equatorial Pacific to stratify, then the natural carbon dioxide that is normally released during venting may just stay in the deep ocean. Stratification might wind up having competing effects on the overall carbon cycle, with saturation slowing carbon dioxide uptake in surface waters, but also suppressing venting.

This is basically saying that due to the destruction of the ozone layer at that time, strong winds and weather were occurring, which were stirring up the ocean and releasing CO2. This showed scientists that when the water was calm, CO2 absorption would decrease, but it would also decrease the amount of CO2 released. The article goes on to say that the Earth (including the ocean) is a very complex environment and there are a lot of factors (both natural and human-made) that affect how it functions.

There have been natural cycles that Earth has been repeating over hundreds of millions of years, but those cycles have been thrown off course by human activity since the Industrial Revolution. Things are becoming more unpredictable, and much more research is required to form a proper model and forecast on what is happening and what will happen in the future to our planet, since historical data won't do us much good anymore.

The oceans are not the only large waterbodies that are undergoing acidification; the Great Lakes and other large lakes around the world are undergoing a similar process. The NOAA has a website for its Acidification program,

Freshwater bodies like the Great Lakes also experience acidification. Researchers project that pH, the measure of how acidic or alkaline the water body is, will decline at a rate like that of the oceans in response to increasing atmospheric carbon dioxide.

Absorption of carbon dioxide from the atmosphere isn't the only source of acidification. The Great Lakes are also recovering from acid deposition. The Midwestern and Northeastern United

States experienced an increase in deposition of sulfuric and nitric acids from the early 20th century until air-quality regulations mitigated this trend.

Present-day mean pH and alkalinity vary according to the geology of each lake basin, with Superior having the most acidified waters and Michigan the least acidified. In addition, considerable short-term spatial and temporal variability in pH occurs, driven largely by varying rates of photosynthesis, respiration, and seasonal mixing. There has not been any long-term robust monitoring of acidification in the Great Lakes system until recently.

CoastAdapt is an Australian organization that provides interested people with information on climate change and responses available to help manage the impacts. They provide the following facts on ocean acidification:

- Oceans absorb a substantial proportion of the CO_2 emitted into the atmosphere by human activities, with potentially negative effects on shell-forming organisms.
- Increasing CO_2 in the atmosphere due to human activities not only affects the climate; it also has direct chemical effects on ocean waters.
- The oceans have absorbed between a third and a half of the CO_2 humans have released into the atmosphere since about 1850. This has slowed the rate of climate change.
- When CO_2 dissolves in seawater, the water becomes more acidic. The acidity of the oceans has increased by 26 % since about 1850, a rate of change roughly 10 times faster than any time in the last 55 million years.
- Associated chemical reactions can make it difficult for marine calcifying organisms, such as coral and some

plankton, to form shells and skeletons, and existing shells become vulnerable to dissolution.

- The extent to which calcifying organisms are already being affected by acidification is unclear, as this is a very new area of study. Limited evidence suggests that some organisms are more sensitive than others.
- The rate at which acidification occurs is a determining factor in the extent to which calcifying organisms will be able to adapt.
- The impacts of acidification will extend up the food chain to affect economic activities such as fisheries, aquaculture, and tourism. Wherever there are marine calcifying organisms, there are risks from ocean acidification.

Below are some of the effects that acidification has on humans:

Changes in marine ecosystems will have consequences for human societies, which depend on the goods and services these ecosystems provide. The implications for society could include substantial revenue declines, loss of employment and livelihoods, and other indirect economic costs.

Socioeconomic impacts associated with the decline of the following ecosystem services are expected:

- **Food:** Ocean acidification has the potential to affect food security. Commercially and ecologically important marine species will be impacted, although they may respond in different ways. Mollusks such as oysters and mussels are among the most sensitive groups. By 2100, the global annual costs of mollusk loss from ocean acidification could be over US$100 billion for a business-as-usual (RCP8.5) CO_2 emissions pathway.

- **Coastal protection:** Marine ecosystems such as coral reefs protect shorelines from the destructive action of storm surges and cyclones, sheltering the only habitable land for several island nations. This protective function of reefs prevents loss of life, property damage, and erosion, and has been valued at US$9 billion per year.
- **Tourism:** This industry could be severely affected by the impacts of ocean acidification on marine ecosystems (e.g., coral reefs). In Australia, the Great Barrier Reef Marine Park attracts about 1.9 million visitors each year and generates more than A$5.4 billion for the Australian economy.
- **Carbon storage and climate regulation:** The capacity of the ocean to absorb CO_2 decreases as ocean acidification increases. More acidic oceans are less effective in moderating climate change.

How We Can Improve the Ocean's Ability to Act as a Carbon Sink

By reducing the amount of CO2 in the atmosphere, we will reduce the amount of carbonic acid being generated by the ocean, which is making it more acidic and creating an inhospitable environment for aquatic life to exist. Since the oceans are the best carbon sinks that we must work with, we need to increase their ability to absorb more CO2 as their ability is decreasing due to becoming too full of CO2, and the CO2 removing aquatic life is dying off. The ocean is a very complex system that is affected by many factors such as currents, weather, pH, water density, and temperature. All these aspects play a role in how effective the oceans are with carbon absorption and carbon venting. As previously mentioned, the mixing of the ocean water can help decrease the amount of absorbed carbon dioxide at the surface of the ocean, bringing in colder, cleaner water from below, which will be able to decrease the acidification process by diluting it. This will make the surface layer of the ocean more hospitable for aquatic life to grow and utilize the CO2 for their photosynthesis. This will also enable the ocean to absorb more CO2 and turn it into carbonic acid and bicarbonate. It is okay for the ocean to create smaller amounts of carbonic acid from CO2; the issue is that it is currently creating too much carbonic acid from the excess CO2 in the atmosphere. By reducing the amount of CO2 in the atmosphere, we will reduce the amount of carbonic acid being generated by the ocean; the catch here is removing the CO2 from the atmosphere since the oceans are the biggest carbon sinks that we must work with, which is why we need to increase their ability to absorb more CO2.

The National Oceanic and Atmospheric Administration of the U.S. Department of Commerce describes ocean currents as follows:

Large-scale surface ocean currents are driven by global wind systems that are fueled by energy from the Sun. These currents transfer heat from the tropics to the polar regions, influencing local and global climate. The warm Gulf Stream , originating in the tropical Caribbean, for instance, carries about 150 times more water than the Amazon River. The current moves along the U.S. East Coast across the Atlantic Ocean towards Europe. The heat from the Gulf Stream keeps much of Northern Europe significantly warmer than other places equally as far north.

Deep ocean currents

Differences in water density, resulting from the variability of water temperature (*thermo*) and salinity (*haline*), also cause ocean currents. This process is known as thermohaline circulation. In cold regions, such as the North Atlantic Ocean, ocean water loses heat to the atmosphere and becomes cold and dense. When ocean water freezes, forming sea ice, salt is left behind, causing the surrounding seawater to become saltier and denser. Dense, cold, salty water sinks to the ocean bottom. Surface water flows in to replace the sinking water, which in turn becomes cold and salty enough to sink. This "starts" the global conveyor belt, a connected system of deep and surface currents that circulate around the globe on a 1000-year time span. This global set of ocean currents is a critical part of Earth's climate system as well as the ocean nutrient and carbon dioxide cycles.

To understand the effects of the wind on ocean currents, we should understand how wind is generated. The National Geographic encyclopedic entry regarding wind states the following:

Wind is the movement of air caused by the uneven heating of the Earth by the Sun. It does not have much substance—you cannot see it or hold it—but you can feel its force. It can dry your clothes in summer and chill you to the bone in winter. It is strong enough to carry sailing ships across the ocean and rip huge trees from the ground. It is the great equalizer of the atmosphere, transporting heat, moisture, pollutants, and dust great distances around the globe. Landforms, processes, and impacts of wind are called Aeolian landforms, such as sand dunes and Loess deposits, which are deposits of silt.

Differences in atmospheric pressure generate winds. At the Equator, the Sun warms the water and land more than it does the rest of the globe. Warm equatorial air rises higher into the atmosphere and migrates toward the poles. This is a low-pressure system. At the same time, cooler, denser air moves over Earth's surface toward the Equator to replace the heated air. This is a high-pressure system. Winds generally blow from high-pressure areas to low-pressure areas.

The boundary between these two areas is called a front. The complex relationships between fronts cause different types of wind and weather patterns.

Prevailing winds are winds that blow from a single direction over a specific area of the Earth. Areas where prevailing winds meet are called convergence zones. Generally, prevailing winds blow east-west rather than north-south. This happens because Earth's rotation generates what is known as the Coriolis effect. The Coriolis effect makes wind systems twist counterclockwise in the Northern Hemisphere and clockwise in the Southern Hemisphere.

The Coriolis effect causes some winds to travel along the edges of the high-pressure and low-pressure systems. These are called geostrophic winds. In 1857, Dutch meteorologist Christoph Buys Ballot formulated a law about geostrophic winds: When you stand with your back to the wind in the Northern Hemisphere, low pressure is always to your left. (In the Southern Hemisphere, low-pressure systems will be on your right.)

When we reviewed the issue of the polar ice caps melting, we discussed how the additional fresh water from the melting ice caps and ocean ice, which contains much less salt concentration, is flowing into the ocean and how the polar ice caps are not forming as they should be. Both issues are slowing down the ocean's natural global conveyor belt, which is reducing the ocean's ability to absorb as much CO2 as it used to, further increasing the amount of CO2 in the atmosphere and making global warming worse. The best way we can help increase the ocean's currents is by preventing the icecaps from melting more and helping them reform. To do this, we need to reduce the land, air, and water temperatures in the North and South poles, and to accomplish that, we need to decrease the amount of solar radiation being absorbed by the ground, help the areas of permafrost reform, which will help to further reduce the regional temperatures. This will also reduce the amount of methane being released into the atmosphere. We have previously discussed how to reduce the amount of solar radiation being absorbed and how to regrow the permafrost. Finally, and probably the most obvious one, we need to decrease the amount of GHGs which we are producing to enable more heat to escape back into outer space That can be done with the help of governments meeting the Paris Agreement goals or creating and sticking to their own aggressive greenhouse gas emissions goals, as well as have everyone utilizing the previously mentioned suggestions in this book.

Additionally, we need to increase the populations of aquatic life that utilize photosynthesis; these include phytoplankton, seagrass, and seaweed. We can create massive seagrass farms along coasts to help remove excess CO2, as well as decrease the amount of chemicals and toxins we dump into the oceans to help create a better environment for aquatic life to flourish.

Changing Ocean Temperature: Winds and Weather

NASA's climate division mentioned the following:

Covering more than 70% of Earth's surface, our global ocean has a very high heat capacity. It has absorbed 90% of the warming that has occurred in recent decades due to increasing GHGs, and the top few meters of the ocean store as much heat as Earth's entire atmosphere.

The effects of ocean warming include sea level rise due to thermal expansion, coral bleaching, accelerated melting of Earth's major ice sheets, intensified hurricanes, and changes in ocean health and biochemistry.

And the U.S. EPA has the following to say about the increasing oceanic temperature and its effects on weather:

- Sea surface temperature increased during the 20th century and continues to rise. From 1901 through 2020, temperatures rose at an average rate of 0.14°F per decade.
- Sea surface temperature has been consistently higher during the past three decades than at any other time since reliable observations began in 1880.
- Based on the historical record, increases in sea surface temperature have largely occurred over two key periods: between 1910 and 1940, and from about 1970 to the present. Sea surface temperature appears to have cooled between 1880 and 1910.
- Changes in sea surface temperature vary regionally. While most parts of the world's oceans have seen

temperature rise, a few areas have experienced cooling, for example, parts of the North Atlantic.

Because the oceans continuously interact with the atmosphere, sea surface temperature can also have profound effects on global climate. Increases in sea surface temperature have led to an increase in the amount of atmospheric water vapor over the oceans.[4] This water vapor feeds weather systems that produce precipitation, increasing the risk of heavy rain and snow (see the Heavy Precipitation and Tropical Cyclone Activity indicators). Changes in sea surface temperature can shift storm tracks, potentially contributing to droughts in some areas.[5] Increases in sea surface temperature are also expected to lengthen the growth season for certain bacteria that can contaminate seafood and cause foodborne illnesses, thereby increasing the risk of health effects.

Cooling the oceans down to safe levels is simple, but not easy. The main thing we need to do is decrease the amount of solar radiation it is absorbing. This has been discussed previously in the book, but just for a refresher, some processes and technologies which are being researched are spreading reflective particles over the ocean to reflect the solar radiation back into the atmosphere to prevent the ocean from absorbing it, injecting chemicals into the top layer of our atmosphere (stratosphere) like sulphur dioxide to help prevent the solar radiation from entering our atmosphere, making clouds closer to the oceans surface more reflective, building a structure in space to significantly reduce the amount of solar radiation entering our atmosphere or creating a layer of material which can orbit the Earth, creating a shield. As mentioned, a cooler ocean is more efficient at absorbing CO2, and by having stronger currents recirculating water deep in the ocean with surface water, the ocean will also be able to absorb CO2 as the new water will have more capacity for CO2 storage.

Aquatic Photosynthesis

Just like on land, the vegetation in the ocean absorbs CO2 and releases oxygen during a process called photosynthesis. The Smithsonian Science Education Centre describes photosynthesis as follows:

The whole process of photosynthesis is a transfer of energy from the Sun to a plant. Plants are called autotrophs because they can use energy from light to synthesize, or make, their own food source. Many people believe they are "feeding" a plant when they put it in soil, water it, or place it outside in the Sun, but none of these things are considered food. Rather, plants use Sunlight, water, and the gases in the air to make glucose, which is a form of sugar that plants need to survive. This process is called photosynthesis and is performed by all plants, algae, and even some microorganisms. To perform photosynthesis, plants need three things: carbon dioxide, water, and Sunlight. Just like you, plants need to take in gases to live. Animals take in gases through a process called respiration. During the respiration process, animals inhale all the gases in the atmosphere, but the only gas that is retained and not immediately exhaled is oxygen. Plants, however, take in and use carbon dioxide gas for photosynthesis. Carbon dioxide enters through tiny holes in a plant's leaves, flowers, branches, stems, and roots. Plants also require water to make their food. Depending on the environment, a plant's access to water will vary. For example, desert plants, like a cactus, have less available water than a Lilypad in a pond, but every photosynthetic organism has some sort of adaptation, or special structure, designed to collect water. For most plants, roots are responsible for absorbing water.

The last requirement for photosynthesis is an important one because it provides the energy to make sugar. How does a plant take carbon dioxide and water molecules and make a food molecule? The Sun! The energy from light causes a chemical reaction that breaks down the molecules of carbon dioxide and water and reorganizes them to make the sugar (glucose) and oxygen gas. After the sugar is produced, it is then broken down by the mitochondria into energy that can be used for growth and repair. The oxygen that is produced is released from the same tiny holes through which the carbon dioxide entered.

Kalila Morsink wrote an article for the Smithsonian National Museum of Natural History, published in July 2017, titled, "With Every Breath You Take, Thank the Ocean." I can't do a better job summarizing what she wrote, so I will just include the entire article below:

When was the last time you thought about your breathing? Take a breath right now and think about it. You breathe because you need oxygen, a gas which makes up 21 percent of the Earth's atmosphere. All that oxygen must come from somewhere. You might already know that it comes from photosynthetic organisms like plants. But did you know that most of the oxygen you breathe comes from organisms in the ocean?

That's right—**more than half of the oxygen you breathe comes from marine photosynthesizers,** like **phytoplankton** and seaweed. Both use carbon dioxide, water, and energy from the Sun to make food for themselves, releasing oxygen in the process. In other words, they photosynthesize. And they do it in the ocean.

Photosynthesizers have been in the ocean for a long time. Land plants started appearing in the **fossil record 470 million years ago**, before dinosaurs roamed the Earth. But the ocean was producing oxygen for billions of years before that. The oldest known fossil is from a marine cyanobacterium, a tiny blue-green photosynthesizer that was releasing oxygen 3.5 billion years ago. In a way, we owe the ocean for all the oxygen that comes from land plants as well, because land plants evolved from green marine algae. If there were a race to put oxygen in the atmosphere, the ocean would have one heck of a head start.

But the ocean's long history of photosynthesis would matter very little to us if not for the photosynthesizers that live in it today. Of these, the most impressive is another cyanobacterium called Prochlorococcus. It is estimated to be more abundant than any other photosynthesizer on the planet, and to be responsible for producing 20 percent of the oxygen in the atmosphere. One in every five breaths you take is owed to Prochlorococcus. Perhaps as amazing is the fact that scientists **only discovered this super-abundant photosynthesizer in 1988**—less than thirty years ago!

Another of the ocean's photosynthesizers is impressive for a different reason. Recall that Sunlight is necessary for photosynthesis. That means all marine photosynthesizers must live in what scientists call the "photic zone"—the layer at the top of the ocean that is illuminated by Sunlight. The **photic zone extends down to about 656 feet** (200 meters) below the surface of the ocean, but it's difficult to put a depth limit on it, because photosynthesizers keep taking photosynthesis down farther than we thought possible. Eight hundred and eighty-six feet below the ocean's surface, you would barely be able to see light. But a type of **red algae called Corallinales nevertheless photosynthesizes at that depth**. The red color of Corallinales comes from a pigment which enables it to absorb blue and green light, which is just about the only kind of light that manages to filter down to the improbable depths at which Corallinales lives. This photosynthesis produces oxygen despite only being able to access the tiniest fraction of Sunlight. Whether they evolved the earliest, photosynthesize the most, or dwell the deepest, marine photosynthesizers are indispensable to our survival. If they weren't around, we wouldn't be breathing. So, occasionally, take a moment to remember the ocean and its photosynthesizers. They help you out with every breath you take.

There is mixed research on whether the changing ocean conditions (becoming more saturated with CO2, becoming more acidic, and warming up) are directly negatively affecting the heavy hitters in photosynthesis, but the main point which continues to come up is how it is creating dead zones in the ocean, which are oxygen-depleted. A research paper was completed by the research for engineering ecology and nonlinear Science, North China Electric Power university, Beijing 102206, China, called "Phytoplankton Community Response to Environmental Factors along a Salinity Gradient in a Seagoing River, Tianjin, China," which explored how different environmental factors affected different species of phytoplankton in the Duliujian River, located in Tianjin City. Some of the factors which were taken into consideration were water temperature, amount of salt in the water (salinity), and pH. Simply put, they found that different species of phytoplankton responded differently in different environments, but temperature and salinity both had negative effects as they were increased.

In an article by an organization called "Grist" titled, "All that plastic in the ocean is a climate change problem too", Clara Manno, a marine ecologist, describes something called the biological carbon pump and how it is being affected by the degrading of tiny pieces of plastics which are also contaminating the oceans, and how the plastic is negatively affecting the ocean's ability to process CO2.

The pump works like this: First, carbon dioxide from the atmosphere dissolves into water at the surface of the ocean. Using photosynthesis, tiny marine algae called phytoplankton then absorb that carbon into their bodies before passing it onto small ocean critters — zooplankton — that eat them. In a final step, zooplankton excrete the carbon as part of "fecal pellets" that sink down the water column. Once these carbon-containing pellets reach the ocean floor, the carbon can be re-mineralized into rocks, preventing it from escaping back into the atmosphere.

The article goes on to discuss that when the phytoplankton ingest the microplastics, zooplankton ingest the phytoplankton. The waste from the zooplankton is more buoyant due to the plastic in the phytoplankton, and it takes much longer for the captured carbon in the zooplankton's waste to reach the bottom of the ocean, which is where it would stay for long-term storage. Instead of sinking to the bottom of the ocean, it could be ingested by other aquatic life and be reintroduced into the environment.

The key to increasing photosynthesis in the ocean is increasing the amount of aquatic life in the ocean, which is responsible for photosynthesis, and that can be done by improving their living conditions to help them grow and flourish. Even genetically altering them to make them more resilient to environments of decreasing pH and increasing temperatures. To increase the pH of the ocean (help it become more neutral), it would have to stop absorbing high quantities of CO2 from the atmosphere and mix with the deeper ocean water, which has a more neutral pH. This won't remove the carbonic acid, but it will at least dilute the acidic waste. This is an issue since the melting ice caps are slowing down the ocean currents, as previously mentioned, but we know how to fix that. Second is to decrease the temperature of the ocean, and that can be done by reducing the amount of solar radiation that is being absorbed by the ocean and reducing the amount of GHGs in the atmosphere, which will let more of the heat escape from our atmosphere; we now know how to do that too. We can decrease the amount of microplastics in the ocean (which will be discussed later) to help increase the phytoplankton population by ensuring they stop eating it and dying off, and increasing the seagrass/seaweed farms all around the world.

The Land

Now, to discuss the carbon sinks you are probably more familiar with, the land-based carbon sinks, which consist of forests, wetlands, and soils. NASA estimates that the forests cover 30% of Earth's surface and store about 45% of the CO2 stored on land.

The following countries have the most vegetation in the world:

1. Russia (815 million Hectares)
2. Brazil (497 million Hectares)
3. Canada (347 million Hectares)
4. United States (310 million Hectares)
5. China (220 million Hectares)
6. Australia (134 million Hectares)
7. Democratic Republican of the Congo (126 million Hectares)
8. Indonesia (92 million Hectares)
9. Peru (72 million Hectares)
10. India (72 million Meters)

The largest forests in the world are as follows:

1. Taiga (1.7 billion hectares) *across both North America and Russia
2. Amazon Rainforest (670 million Hectares)
3. Congo Rainforest (200 million hectares)
4. Tongass National Forest (93 million hectares)
5. Forests of New Guinea (73 million hectares)
6. Valdivian Rainforest (25 million Hectares)
7. Sundarbans (1 million hectares)
8. Rainforest of Xishuangbanna (240 thousand hectares)
9. Daintree Forest (120 thousand Hectares)
10. Kinabalu National Park (75 thousand Hectares)

In case you were wondering what the difference is between a jungle and a forest, all jungles are forests, but not all forests are jungles. Per National Geographic, a forest is the following: "The word forest broadly describes an area that has many trees. There are three general types of forests that exist: temperate, tropical, and boreal. Experts estimate that these forests cover approximately one-third of Earth's surface." The Britannica dictionary describes a jungle as the following: "The word 'jungle' is usually used to describe a tangled or overgrown mass of vegetation over a large area of land. A jungle usually has a tropical or humid climate and many plants on the ground between trees and larger plants."

Different types of trees have different roles and abilities regarding absorbing CO_2. The age and health of the trees also play a role. The tree's ability to thrive also depends on its climate and environment. Healthy trees in their right zone will do a much better job of absorbing CO_2 than the same trees in a harmful environment. The higher the amount of biodiversity an area has (the more different types of trees and other vegetation), the better an area will be at absorbing CO2. Ross Bernet wrote an article for One Tree Planted called "How much CO2 does a tree absorb," published on July 25, 2023, and he states that the main factors that affect tree growth are as follows:

Location

Growing Conditions

Water Availability

Sunlight

Local Climate

Soil Nutrients

Site Specific Factors

Avely Putsep wrote an article called "Which Trees Absorb the most Carbon and Why?" for an organization called Single. Earth. It states that trees with large trunks and dense wood tend to be the best absorbers, and large leaves and wide crowns help too.

Other features that help a tree absorb carbon include:

- **Being fast-growing:** Most sources say trees store most carbon during their first decades
- **Being native species:** This helps them thrive in the soil.
- **Being low-maintenance:** They won't need fertilizers and other chemicals

It is difficult to say which trees are the best at removing CO2 from the air, as it would differ from region to region, but generally, pines, oaks, and mangroves are known to absorb the most carbon.

Mangroves may just be the best trees when it comes to fighting climate change for the following reasons:

Their benefits include:

Storing massive amounts of carbon in trees and the surrounding soil

- Reducing flooding and erosion from storms
- Acting as nurseries for fish
- Filtering pollutants from water

An article was published on May 6, 2022, and was written by Ralph J.M. Temmink et al, titled "Recovering wetland biogeomorphic feedbacks to restore the world's biotic carbon hotspots." In this publication, they state the following,

Although biogeomorphic wetlands (peatlands, mangroves, salt marshes, and seagrass meadows) cover only 1% of Earth's surface, they store 20% of the global organic ecosystem carbon. This disproportionate share is fueled by high carbon sequestration rates per unit area and effective storage capacity, which greatly exceed those of oceanic and forest ecosystems.

An article prepared by Frederic Forge in the Science and Technology Division of the Government of Canada (published January 30, 2001) mentions the following regarding CO_2 production in soils, specifically farmland:

To understand CO_2 emissions from soil, it is important to have some idea of the carbon cycle in farm systems. Generally, CO_2 in the atmosphere is absorbed by plants, which transform it into carbohydrates, cellulose, and other sugars. Each plant uses some of the carbon compounds to meet its energy needs and converts them back into CO_2. Some of the carbon remaining in the plant is then removed from the system when the plant is harvested; the rest ends up in the ground and is transformed into CO2 again by microbes in the soil. This cycle is identical in all crop systems, but the quantities of CO2 involved vary depending on climate, soil, and type of plant. The cycle is slightly more complex on farms where animals are raised because, instead of being removed from the system, a considerable proportion of the plant matter is used as bedding or feed. Some of that carbon is released by the animals in the form of CO2, some is removed in the form of animal products (meat, for example), and a significant quantity is returned to the ground in the form of manure.

It is not only soil that can store CO2, but rocks do as well. An article published by the Yale School of the Environment, E360 Digest on August 15, 2023, titled "Spreading Rock Dust on Farmland has potential to draw down huge sums of Carbon Dioxide," mentions that rain absorbs carbon from the air as it falls, and it reacts with rocks on the ground to lock away carbon.

Loss of Land-Based Carbon Sinks

We have lost a lot of land-based carbon sinks over this past century in the form of trees, forests, and wetlands. Martina Igini wrote an article for Earth.org titled "10 Shocking Statistics About Deforestation," and it was published on January 21, 2023. She mentioned that since 1990, we have lost 420 million hectares of forest across the world. Below are the ten facts she mentions; she goes into more detail in the article, you just must look it up.

1. We lose approximately 10 million hectares a year.
2. We are losing forests at a rate equivalent to 27 soccer fields per minute.
3. 95% of deforestation happens in the tropics.

4. 99% of Haiti's native forests have been cut down.
5. 75% of Tropical Rainforests have lost resilience.
6. Between 2017 and 2015, Tropical Forests released 10 billion tonnes of carbon dioxide.
7. Nearly 1.6 billion Hectares of Forest Are at High Risk of Degradation.
8. Due to forest conversion, Luxembourg's emissions reached 2.8 tonnes per year, the highest in the world.
9. 25% of Western drugs used to treat diseases come from the rainforest
10. Over one million Indigenous people, as well as thousands of animal and plant species, are threatened by deforestation.

I spoke earlier about an argument I commonly get into regarding cutting down old growth trees for construction material and to create a healthier forest. One side of the argument is that the old-growth trees are part of a delicate ecosystem. All of the vegetation is connected to the surrounding vegetation through their roots and a fungus called mycelium which helps the trees and other vegetation communicate and share water and nutrients; the old-growth trees are also home to wildlife. It is a great argument, and it is all true.

The other side of the argument is that the older trees are not as good at removing carbon as younger trees, and we can do a better job with long-term carbon storage by selectively cutting down the old growth trees, using them for lumber and locking the absorbed carbon up in our homes and buildings instead of letting it just be released as the tree naturally dies and decomposes. The National Council for Air and Stream Improvement Inc says the following regarding old grown vs young growth forests:

Forests of different ages play different roles in removing carbon from the atmosphere and storing it in wood. Old forests have accumulated more carbon than younger forests; however, young forests grow rapidly, removing much more CO2 each year from the atmosphere than older forests covering the same area. Managing forests to avoid large emissions from the loss of old trees while rapidly removing CO2 from the atmosphere through young forest growth can provide both storage and sequestration benefits. In addition, managed forests produce wood products that store carbon long after the trees are harvested. These products provide an added benefit when they are used in place of more energy-intensive ones that require more fossil fuel emissions.

It is tough to say who is right and wrong in that argument. It really depends on what the final goal is; is it to remove and store as much CO2 as possible to help reduce extreme climate change, or is it to not disrupt the local ecosystem? I would argue that by doing everything we can to reduce the CO2 in the air, locking it in and making sure it does not escape to help reduce the CO2 concentrations in the atmosphere will benefit both the global and local ecosystem, but it will be more beneficial to the global warming issue, which affects everyone and everything. I am not suggesting we need to go and start clear-cutting old growth forests, but I do think the best choice is selective logging, which identifies specific trees that have done their part in capturing carbon, cuts them down, and removes them for processing before they can naturally die and release the stored carbon. We will also need to replace the trees with younger trees so they can do a more efficient job at removing CO2 from the air and help save the planet. I understand that this could temporarily disrupt the ecosystem, but looking at the bigger picture, I think the pros outweigh the cons. Regarding wetlands, I came across a publication called "Global Wetland Outlook – Special Edition 2021" from the Convention on Wetlands, and if the current conditions of the global wetlands interest you, I would recommend this publication. A few takeaways from this I would like to share here are as follows:

The first Global Wetlands Outlook was published in 2018 (Ramsar Convention, 2018a). It reported that although wetlands still cover a global area of 1.2 billion hectares (ha) – larger than Canada – they are declining fast, with 35% losses of natural wetlands since 1970, where data are available. The quality of remaining wetlands is also suffering due to drainage, pollution, invasive species, unsustainable use, disrupted flow regimes, and climate change. Yet, wetland ecosystem services, ranging from food security to climate change mitigation, remain enormous, far outweighing those of terrestrial ecosystems.

1) Deterioration of wetlands is increasingly widespread, although more wetlands are reported as currently being in good rather than bad ecological character.
2) Biodiversity losses are continuing to rise, but are also better understood than previously.
3) Multiple sources of pollution, particularly from agriculture, continue to degrade wetland ecosystems.
4) Improvement in wetland ecological character is found to be linked to the implementation of the Convention on Wetlands Strategic Plan commitments.
5) Wetlands are particularly impacted by climate change, especially sea-level rise, coral bleaching as a result of increased sea surface temperatures, and changing hydrology in inland waters.
6) Arctic and montane wetlands are especially at risk.
7) Changing weather patterns also increase risks of water stress – both flooding and drought – in many parts of the world.

8) New information on the rate of climate change emphasises the urgency of wetland management and restoration for both mitigation and adaptation.

With the major losses of forests, biodiversity, and wetlands, we are decreasing the Earth's natural ability to remove CO2 from the air. We need to change how we function with regard to land development and agriculture, as well as investing in major restoration and regeneration projects to recover the wetlands and forests to get back our natural, low-maintenance, relatively inexpensive carbon sinks. It is important to note that every carbon sink can become a carbon source. The carbon sinks do change some CO2 into different forms, but also store the CO2. As the wetlands and trees die and decompose or are burnt through fires (natural or human-made), or when the soil quality reduces, they all release CO2 and become carbon sources.

Reasons for Land-Based Carbon Sink Loss

Human Causes

Agriculture is the main reason for the extensive deforestation and draining of wetlands around the world. According to the World Wildlife Federation, the two largest products requiring the most land to be converted to agricultural land are Beef and Soy; between 70% and 75% of all soy becomes livestock feed for chickens, pigs, farmed fish, and cows.

Beef production is the top driver of deforestation in the world's tropical forests. The destruction of the Amazon Rainforest specifically for beef production is more than double what is caused by the production of soy, palm oil, and wood products (the second, third, and fourth biggest drivers) combined. Beef production also drives the conversion of non-forest landscapes from grasslands to savannas through overgrazing.

The second biggest driver of tropical deforestation is soybean production. Global soybean production has increased more than fifteenfold since the 1950s. Between 1990 and 2010, in South America, the land used for soy production swelled from 42 million acres (roughly the size of Washington State) to 114 million acres (bigger than the state of California). By reducing the amount of cattle we have to feed, we will reduce the amount of feed we need to produce, and we will not need to use as much of the land for farming soy or other feed.

According to the EPA, in the U.S.

More than 220 million acres of wetlands are thought to have existed in the lower 48 states in the 1600s. Since then, extensive losses have occurred, and more than half of our original wetlands have been drained and converted to other uses. The mid-1950s to the mid-1970s were a time of major national wetland loss. Since then, the rate of loss has slowed.

Pollution Inputs. Although wetlands are capable of absorbing pollutants from surface water, there is a limit to their capacity to do so. The primary pollutants causing wetland degradation are sediment, fertilizer, human sewage, animal waste, road salts, pesticides, heavy metals, and selenium. Pollutants can originate from many sources, including runoff from urban, agricultural, silvicultural, and mining areas. Air pollution from cars, factories, and power plants. Old landfills and dumps that leak toxic substances. Marinas, where boats increase turbidity and release pollutants.

The UN mentions that, "the world's remaining wetlands are under threat due to water drainage, pollution, unsustainable use, invasive species, disrupted flows from dams and sediment dumping from deforestation and soil erosion upstream."

Natural Causes

Droughts and wildfires are major natural occurrences that have destroyed large amounts of forest and wetlands. There have also been increasing invasive plants, insects, and diseases that have been ravaging the forest, killing all the trees, causing them to start decomposing and release all the CO2 they have absorbed throughout their lifespan.

How to improve the forest carbon sink

The obvious answer here is to stop the massive deforestation happening all over the world, which can be accomplished by decreasing the demand for what the forest is being destroyed for. The second obvious answer here is for us to plant more trees and help the forests grow instead of shrinking. The third best way to help improve the natural forest carbon sinks is to prevent massive wildfires, which are getting worse by the year.

In Vancouver, British Columbia, Canada, where I currently am, it is only March 2024, and there is already an air quality advisory due to the wildfires. It was only a decade ago that I could go camping, or up to that resort I mentioned earlier, and have clean air, no fire bans, and be able to really get out there and enjoy nature. The past summers have been horrible regarding wildfires, with parts of major cities being destroyed, electrical transmission towers and lines being destroyed, and potentially cutting regions off from power. Additionally, a very environmentally hazardous chemical used in electrical equipment called Polychlorinated biphenols (PCBs) has spilled into the environment due to the wildfires. This chemical is highly detrimental to the environment. The Canadian federal government has ordered that all oil with over 50 parts per million of PCBs needs to be removed by 2025.

Wildfires are horrible for many reasons, including loss of lives, homes, businesses, increasing pollution, ruining landscapes and habitats, decreasing air quality, and overall quality of life from the smoke. Wildfires are also costing governments hundreds of millions of dollars of taxpayers' money, but for the sake of this book, we will just focus on the addition of CO2 and the decreasing amount of trees that could have helped reduce the CO2 concentrations.

Forest management plays a large part in reducing wildfires. Small wildfires are natural and beneficial for the environment as they help to clear out the dead material on the forest floor, release nutrients back into the soil, and open up the canopy in areas to let the Sunlight through; all creating an ideal environment for new growth. A healthy forest needs to have old growth, new growth, and dead material, which is decomposing; this is how the planet has survived for millions of years. Due to the massive change in population, cities, roads, and land required for agriculture, we have built our infrastructure in areas that were previously covered by forest. The wildfire issues have many aspects, which are both human-made and natural. With the heating up of the Earth and less precipitation where it used to fall, the forests are drying up, and the winters are becoming warmer (killing off less destructive insects in some areas). In British Columbia, between April 1, 2021, and March 30, 2022, there were 1,642 wildfires, 869,279 hectares were burnt, and approximately 60% were caused naturally from things like lightning strikes, while 35% were caused by humans, and the remaining 5% had unknown causes. There is more human interaction with the forests which includes camping fires or disposing of burning cigarette butts which can ignite wildfires, also we are putting out some fires a lot quicker instead of letting them burn since they are close to rural areas; preventing some wildfires just causes the dead material on the forest floor to build up and create a perfect environment for a massive wildlife. Proper forest management is very important, which is why we need experts in this field as well as to offer education to people who are looking to do their part with wildfire prevention. Forest management is crucial in preventing the massive wildfires we have been experiencing over the past decade and will continue to experience in the near future until a major change occurs. Forest management is vast and includes more then just cleaning the dead material off of the forest floor, but other things that we can control which will play a mayor role include removing the older / dying tress, planting new more lively trees and vegetation, increasing the diversity of the vegetation in the forests, preventing the climate from changing too much so the forest is able to receive the appropriate amount of precipitation and preventing it from becoming dried out. Cloud seeding could also help keep the forests wet, help promote growth, and reduce the risk of wildfires.

FireSmart Canada developed the Firesmart program, which "helps reduce wildfire risks to homes, neighbourhoods, critical infrastructure, and vital natural resources. This is a shared responsibility between us all. Individuals, communities, governments, and private sector organizations all have a part to play." It consists of seven areas: education, emergency planning, vegetation management, legislation, development, interagency cooperation, and cross training. To learn more about the program and how to better protect your home, business, and community from wildfires, please visit their website, but I will add some notes in here to give you a general understanding.

Education: Through online and printed resources available from FireSmart Canada, education can be promoted and shared within a neighbourhood, raising further interest and awareness among more residents.

Emergency Planning: There are two ways of approaching the planning discipline: the creation of wildfire prevention and mitigation specific plans, and accounting for wildfires in existing emergency response and neighbourhood preparedness.

Vegetation Management: Fuel Removal – flammable vegetation is removed. For example, removing mature spruce trees adjacent to one's home. Fuel Reduction – flammable vegetation is lessened; examples include pruning, thinning, mowing, grazing, mulching, hazard reduction burning, or pile burning. Species Conversion – removal of flammable species and replacement with less flammable species. An example is replacing conifers with deciduous species. Reducing, removing and converting combustible vegetation within the home ignition zone on our private properties is critically important. More than 90% of homes damaged or destroyed by wildfires are ignited by embers. Maintaining a 1.5 metre non-combustible surface around your entire home will reduce the chance of wind-blown embers igniting materials near your home.

Legislation - Post-fire studies, experiments and models have shown that homes ignite due to the condition of the structure and everything around it, up to 100 metres from the foundation. Legislation, bylaws, plans, policies and development standards all play a significant role in building and maintaining FireSmart neighbourhoods. Consideration of wildfires at the development planning stage is a key step in protecting neighbourhoods from wildfires.

Development - Development standards play a significant role in reducing the potential impact a wildfire may ultimately have on a neighbourhood. The potential for damage intensifies when flammable building materials are used.

Interagency Cooperation - A successful neighbourhood FireSmart program requires engagement and strong partnerships. Elected officials, neighbourhood planners, developers, government, industry, and residents all have an important role to play.

Cross Training - Cross-training aims to bring emergency response organizations together for the purpose of identifying potential weaknesses and gaps in response efforts prior to an actual emergency.

As I just mentioned, natural causes are a major contributor to wildfires, and the main natural cause is lightning. There is a strong argument for how extreme climate change is increasing the intensity and frequency of storms, which cause lightning, which would increase the amount of natural wildfires.

Parks Canada has the following to say about wildfires.

As our climate changes, fire seasons are predicted to become longer. Wildfires may become larger and burn more severely. This is partly due to the expected drought that will create dry, flammable conditions across much of the country. Reintroducing fire can create more diverse landscapes and improve the ecosystem's ability to adapt to the changing climate. Prescribed fire and wildfire management help accomplish this. Resilient landscapes mean that the forests, grasslands, and other ecosystems can recover and persist, even with the predicted impacts of climate change.

In 2023, Parks Canada dealt with 109 wildfires in 20 different parks and sites which destroyed over 1 million hectares of forest, which means that these once carbon sinks turned into massive carbon sources, as well as destroyed lives, homes, businesses, infrastructure and cost the taxpayers hundreds of millions of dollars.

Forest management is the key to the prevention of wildfires by creating smaller fires to help prevent the massive fires, but the issue is the total area of the forests and the limited resources available. As the governments are spending hundreds of millions of dollars a year on wildfires, they must have a lot of interest in new technologies that are less expensive and help prevent the damage caused by wildfires. These technologies could be for either early detection of wildfires so they can put out before they get out of control, or something that will help extinguish the wildfires better than the water bombers; something like portable towers/silos filled with a fire suppression chemical or some sort of a projectile that could be launched from the ground or dropped from a airplane which can explode in the air and release the fire suppression chemicals over a wide area.

Land Clearing

Land clearing is the process of clearing land for operations like constructing buildings, parking lots, farms, mines, etc. This is a large cause of deforestation globally. For example, the recent demand for avocados has caused massive amounts of land clearing for new avocado farms, all this just so you can have your avocado toast every morning. The World Economic Forum mentions the following:

"Forest lands with diverse wildlife have been destroyed to produce avocado, and many more were intentionally burned to bypass a Mexican law allowing producers to change the land-use permit to commercial agriculture instead of forest land, if it was lost to burning. A study conducted by Carbon Footprint Ltd affirms a small pack of two avocados has an emissions footprint of 846.36g CO_2, almost twice the size of one kilo of bananas (480g CO_2) and three times the size of a large cappuccino with regular cow milk (235g CO_2)."

An easy thing you can do to help reduce the amount of land clearing required to generate your food is to spend two minutes and lookup what fruits and vegetables are grown locally near you and are in season, then do your best to incorporate them into your meals and not put a demand on the fruits and vegetables that are imported from around the world and most likely not produced in a sustainable way. This will not only help the environment, but it should also help support your local farmers, which is a major win-win.

As mentioned before, beef and soy crops are the top two reasons for land clearing, and the third is palm oil. Palm oil is found in food, cosmetics, detergents, soap, and is even used as a biofuel. Rainforest Rescue said the following about palm oil: Day after day, huge tracts of rainforest in Southeast Asia, Latin America, and Africa are being bulldozed or torched to make room for more plantations, releasing vast amounts of carbon into the atmosphere. Therefore, Indonesia – the world's largest producer of palm oil – temporarily surpassed the United States in terms of greenhouse gas emissions in 2015. With their CO_2 and methane emissions, palm oil-based biofuels have three times the climate impact of traditional fossil fuels. They recommend the following steps to help you do your part to combat the production of palm oil

1. **Enjoy a home-cooked meal: Use your imagination:** why not try almond-coconut-pear biscuits? Or pizza with potato and rosemary? A meal cooked from fresh ingredients beats processed foods containing palm oil every time. Oils such as Sunflower, olive, rapeseed, or flaxseed are ideal for cooking and baking.
2. **Read labels**: As of December 2014, labeling regulations in the EU require food products to clearly indicate that they contain palm oil. However, in the case of non-food items such as cosmetics and cleaning products, a wide range of chemical names may still be used to hide the use of palm oil. A quick check of your favorite search engine will turn up palm oil-free alternatives, however.
3. **Remember that the customer is king:** Ask your retailers for palm oil-free products. Write product manufacturers and ask them why they aren't using domestic oils. Companies can be quite sensitive to issues that give their products a bad name, so inquiring with sales staff and contacting manufacturers can make a real difference. Public pressure and increased awareness of the problem have already prompted some producers to stop using palm oil.
4. **Sign petitions and write your elected representatives:** Online campaigns put pressure on policymakers responsible for biofuels and palm oil imports. Have you already signed all of Rainforest Rescue's petitions?
5. **Speak out:** Protest marches and creative action on the street raise public and media awareness of the issue, which in turn steps up the pressure on policymakers.
6. **Leave your car at home**: Whenever you can, walk, ride a bicycle or use public transport.
7. **Be informed and inform others:** Big Business and governments would like us to believe that biofuels are good for the climate and that oil palm plantations are

sustainable. Spread the word – share this information with your family and friends and encourage them to rethink their consumption habits. It's in our hands!

Another way we can help reduce the production of palm oil and reduce the amount of land clearing is by making it illegal to import and use palm oil containing products. This has been done with many products in the past which were found to have negative effects on the consumers' health and the environment; recently, the US EPA banned the use of a type of asbestos. Other similar things that have been banned over time would be consumer products containing mercury, lead, PCBs, CFCs, asbestos etc. There is a clear connection between the mass land clearing and the production of beef, soy and palm oil products; unlike meat and soy, the oil products can easily be replaced with something else and the world would still function normally without it, so why don't we try it? To stop the massive land clearing, we must stop the demand for the products that are causing the land clearing. If there is no demand, there would be no land clearing.

Forestry

Proper logging processes can make a huge difference in maintaining a healthy forest, preventing massive out-of-control wildfires, and keeping the forests acting as a strong carbon sink. Through processes like selective logging and replanting a variety of local vegetation to help with bio-diversity of the region, we will manage to meet our demand for the forestry products we require while maintaining a healthy forest ecosystem. The Rainforest Alliance says the following about sustainable forestry,

For many, the term "sustainable forestry" may seem like an oxymoron. How can logging ever be sustainable when, by definition, it requires that trees be cut down? The complete answer is complex, but if we had to reduce it to one word, it would be "balance." The hallmark of sustainable forestry, from a purely ecological perspective, is the extent to which forestry practices mimic natural patterns of disturbance and regeneration. Sustainable forestry balances the needs of the environment, wildlife, and forest communities, supporting decent incomes while conserving our forests for future generations. There are many practical steps that a community or business can take to protect the health and longevity of a forest while still profiting from the production and sale of timber and other forest products, such as nuts, fruits, oils, and plants.

In 1993, the Rainforest Alliance created the Forest Stewardship Council (FSC). This Council is an international body that develops and maintains the most globally respected standard for managing forests, and it certifies forestry businesses that meet the standard's strict environmental, social, and economic criteria. They also award FSC Chain-of-Custody (CoC) certificates for the tracking of certified wood and other raw materials from forest to manufacturer to store shelf, confirming that an item that bears the FSC and Rainforest Alliance Certified trust marks contains sustainable products. These seals provide conscientious consumers an easy way to spot and support responsible producers; just look for their tree check marking. For more information on how you can be a responsible consumer regarding supporting sustainable forestry, please visit their website.

Alternative Forestry Methods

There is a Japanese forestry method which stands out in my head as something with a lot of potential to create sustainable logging; it is called Daisugi, which apparently literally translates into platform cedar. Grace Ebert published an article on October

22, 2020, titled "A Japanese Forestry Technique Prunes Upper Branches to Create a Tree Platform for More Sustainable Harvests," and the Daisugi method is described as follows:

The method originated in Kyoto and involves pruning the branches of Kitayama cedar so that the remaining shoots grow straight upward from a platform. Rather than harvesting the entire tree for lumber, loggers can fell just the upper portions, leaving the base and root structure intact.

Although daisugi mostly is used in gardens or bonsai today, it originally was developed to combat a seedling shortage when the demand for taruki, a type of impeccably straight and knot-free lumber, was high. Because the upper shoots of Kitayama cedar can be felled every 20 years, which is far sooner than with other methods, the technique grew in popularity.

By logging this way, we would not be killing the tree, harming the ecosystem and the underground fungal layer, as it would simply be removing "branches" and letting them grow back.

Agroforestry

The USDA describes agroforestry as follows: Agroforestry is the intentional integration of trees and shrubs into crop and animal farming systems to create environmental, economic, and social benefits. It has been practiced in the United States and around the world for centuries.

There are five main types of Agroforestry:

Agroforestry Farming Systems

Alley cropping means planting crops between rows of trees to provide income while the trees mature. The system can be designed to produce fruits, vegetables, grains, flowers, herbs, bioenergy feedstocks, and more. This type of system may also be called intercropping, when the trees and crops are not in defined rows and alleys.

Forest farming operations grow food, herbal, botanical, or decorative crops under a forest canopy that is managed to provide ideal shade levels as well as other products. Forest farming is also called multi-story cropping.

Silvopasture combines trees with livestock and forage on one piece of land. The trees may provide timber, fruit, fodder, or nuts as well as shade and shelter for livestock and their forages, reducing stress on the animals from the hot summer Sun, cold winter winds, or a downpour.

Linear Agroforestry Practices

Riparian forest buffers are natural or re-established areas along rivers and streams made up of trees, shrubs, and grasses. These buffers can help filter farm runoff while the roots stabilize the banks of streams, rivers, lakes and ponds to prevent erosion.

These areas can also support wildlife and provide another source of income.

Windbreaks shelter crops, animals, buildings, and soil from wind, snow, dust, and odors. These areas can also support wildlife and provide another source of income. They are also called shelterbelts, hedgerows, vegetated environmental buffers, or living snow fences.

An article written by Sarah E Castle, Daniel C Miller, Nikolas Merten, Pablo J Ordonez and Kathy Baylis called "Evidence for the impacts of agroforestry on ecosystem services and human well-being in high-income countries: a systemic map." Published on October 29, 2018, states that Agroforestry is promoted for its potential for carbon capture, soil erosion and runoff control, and improved nutrient and water cycling, as well as for offering socio-economic benefits and greater agricultural productivity. The potential for agroforestry to provide sustainable production, improve food security, increase water quality, combat climate change and biodiversity loss, and reduce poverty also provides an opportunity to advance the UN 2030 Sustainable Development Goals.

The biodiversity created from agroforestry helps not only generate a healthier ecosystem, but also increases a farm's ability to capture carbon and increase its ability to act as a carbon sink.

Rocks

We touched on cover crops earlier, so I won't get too much into details with it, but basically, between the cash crops, if farmers plant a cover crop, it can increase the soil quality, help produce a higher yield, help prevent soil run off and help with carbon capture and carbon storage. A newer method that can help farmers with soil quality, maintain moisture in the soil and help remove a lot more carbon from the air is by spreading certain types of crushed rocks over their fields. Susan Cosier published an article on September 2, 2021, for Yale Environmental 360 titled, "How adding Rock Dust to Soil Could Help Get Carbon into the Ground," and it describes how pulverized silicate rocks should be applied to agricultural fields to not only improve the crops, but also remove far more amounts of CO2 from the atmosphere. Researchers are spreading basalt on cornfields in Illinois, Sugarcane in Australia and Soybean fields in Canada. According to the U.N. Intergovernmental Panel on Climate Change (IPCC), rocks naturally remove 1 gigaton (1 billion tons) of carbon dioxide a year from the atmosphere (a number that has changed over time). Adding rock dust to agricultural lands speeds up the chemical reactions that lock carbon up — for thousands of years — in soil. If applied to croplands globally, rock dust could theoretically help suck an estimated 2 to 4 billion tons of carbon dioxide from the air every year, between 34 and 68 percent of the global greenhouse gas emissions produced by agriculture annually. This is a great opportunity to set up a circular economy and see if we can utilize some of the waste rock from mining operations or drilling and blasting from road and infrastructure development to be used to spread over fields and even barren landscapes to help remove more carbon from the air than it would have been normally.

Restoring the Wetlands

There are pros and cons to restoring or constructing wetlands. In an article by P.E.R. Dale and R. Connelly, published June 12, 2012, titled "Wetlands and Human Health: an overview", it mentions the following:

In terms of human health, wetlands have both positive and negative impacts. This was the theme of a Ramsar meeting in 2008 (Bridgewater 2008) and has recently been stressed by Horwitz and Finlayson (2011). Wetlands contribute to mental and social well-being, which are positive aspects of health. Whitehead and Rose (2009) recently estimated the environmental benefits of a range of natural processes, including wetland contributions to recreation, an aspect of well-being. There are mental health benefits of experiencing the natural environment and its therapeutic value, including 'ecotherapy', exemplified by the recent paper by Stevens (2010). Abraham et al. (2010) reviewed over 120 studies related to well-being and outdoor environments and the health-promoting impact of landscape. While these are not necessarily focused on wetlands, they have application to these systems. Social well-being includes human well-being generally, as recently noted by Horwitz and Finlayson(2011), who provided a balanced overview and model for managing wetlands to conserve ecosystem services and thus protect human health. The physical health dimension perceived with wetlands may be a negative one, with a range of disease risks. Of the disease vectors, mosquito-borne disease is the most often reported in the literature and was reviewed in Dale and Knight (2008). Negative health impacts may also be associated with constructed wetlands or wetlands polluted by pathogens, heavy metals or pesticides, which pose a health risk particularly when human populations and their activities encroach on the habitats. Thus Graczyk et al.(2009) found human enteropathogens (Cryptosporidium spp.) associated with constructed wetlands in Ireland; Sekomo et al. (2011) recently found concentrations of a variety of heavy metals in natural wetlands receiving wastewater in Rwanda which posed a health hazard for local people. Parasites may be found in natural wetlands, especially in those that are degraded (Shapiro et al. 2010).

Wetlands are crucial for removing and storing CO2 in the atmosphere, can be low maintenance, and create an ecosystem for life to thrive, but as just mentioned in the article above, they can cause serious health problems for humans. Wetlands also have many positive impacts on the environment other than their ability to act as a powerful carbon sink. According to the EPA,

Wetlands can play a role in reducing the frequency and intensity of floods by acting as natural buffers, soaking up and storing a significant amount of floodwater. A wetland can typically store about three acre-feet (three acres covered in water one foot deep) or about one million gallons of water per acre of wetland. Coastal wetlands serve as storm surge protectors when hurricanes or tropical storms come ashore. In the Gulf coast area, barrier islands, shoals, marshes, forested wetlands and other features of the coastal landscape can provide a significant and potentially sustainable buffer from wind wave action and storm surge generated by tropical storms and hurricanes.

Other benefits that the EPA discusses are:

- Wetlands provide critical habitat, breeding grounds and sources of food for shellfish, fish, birds, amphibians, and other organisms.
- Wetlands play a crucial role in many state and tribal fishing economies.
- Wetlands are also preserved to provide feeding and resting grounds for migratory birds and to create habitat corridors for wildlife populations.
- These services generate state and tribal commercial, recreational and aesthetic benefits as well.
- Wetlands also control erosion, limit flooding, moderate groundwater levels and base flow, assimilate nutrients,

protect drinking water sources and buffer coastal areas from storm surges.

- States may pursue wetland restoration to improve water quality and comply with Total Maximum Daily Load (TMDL) pollutant allocations in impaired waters and watersheds.

If you are interested in restoring wetlands in your area, do some research first to see if there are grants being offered in your area. In the U.S., the EPA provides a list of grants (free money) they and other federal agencies are willing to give out for wetland restoration projects; I know Canada also offers grants for environmental projects like wetland restoration. If you are planning on restoring wetlands, the EPA has provided a lot of information on how to do it, and I will provide some of the information below:

Restore ecological integrity. Restoration should re-establish, insofar as possible, the ecological integrity of degraded aquatic ecosystems. Ecological integrity refers to the condition of an ecosystem -- particularly the structure, composition, and natural processes of its biotic communities and physical environment.

An ecosystem with integrity is a resilient and self-sustaining natural system able to accommodate stress and change. Its key ecosystem processes, such as nutrient cycles, succession, water levels and flow patterns, and the dynamics of sediment erosion and deposition, are functioning properly within the natural range of variability. Biologically, its plant and animal communities are good examples of the native communities and diversity found in the region. Structurally, physical features such as the dimensions of its stream channels are dynamically stable.

Restore natural structure. Many aquatic resources in need of restoration have problems that originated with alteration of channel form or other physical characteristics, which in turn may have led to habitat degradation, changes in flow regimes and siltation. Stream channelization, ditching in wetlands, disconnection from adjacent ecosystems, and shoreline modifications are examples of structural alterations that may need to be addressed in a restoration project. In such cases, restoring the original site's physical attributes is essential to the success of other aspects of the project, such as improving water quality and bringing back native biota.

Restore natural function. Structure and function are closely linked in river corridors, lakes, wetlands, estuaries and other aquatic resources. Reestablishing the appropriate natural structure can bring back beneficial functions.

Use natural fixes and bioengineering techniques, where possible. Bioengineering is a method of construction combining live plants with dead plants or inorganic materials, to produce living, functioning systems to prevent erosion, control sediment and other pollutants, and provide habitat. Bioengineering techniques can often be successful for erosion control and bank stabilization, flood mitigation, and even water treatment. Specific projects can range from the creation of wetland systems for the treatment of storm water, to the restoration of vegetation on river banks to enhance natural decontamination of runoff before it enters the river.

The EPA also mentioned the following regarding wetland restoration:

1) Understand the natural potential of the watershed.
2) Address ongoing causes of degradation.

3) Develop clear, achievable and measurable goals.
4) Focus on feasibility.
5) Use a reference site.
6) Anticipate future changes.
7) Involve the skills and insights of a multi-disciplinary team.
8) Design for self-sustainability.
9) Use passive restoration, when appropriate.
10) Restore native species and avoid non-native species.
11) Monitor and adapt where changes are necessary.

There is money sitting there for you to use if you are interested in helping fight global warming in a big way, and with properly conducted research, you will be able to teach yourself to successfully restore degraded wetlands and make a major positive change in the environment. By doing that, you would be bringing back a thriving ecosystem, building a powerful carbon sink and helping with all the other previously mentioned positive environmental impacts which come along with a healthy wetland.

Waste Management

I broke up this book into two major sections, Extreme Climate Change and Waste Management, because I want to show that there are two major distinct issues that we are dealing with regarding the health of the planet. There is some cross over between Waste Management and Extreme Climate change like how the tiny pieces of plastic in the ocean are killing off many phytoplankton and reducing the ocean's ability to process CO2, when food waste (organic waste) is disposed of at a landfill which does not have a methane capture system and leaks methane into the environment and something like not by not disposing of hazardous waste properly. Saving a turtle's life by properly disposing of a plastic straw or plastic bag will not affect the extreme climate change issue we are facing at all, but it still is a good thing to do. How changing from plastic to paper straws would affect climate change is through the manufacturing process. Manufacturing a paper straw requires less energy and more environmentally friendly raw materials than a plastic straw requires. The process of creating a plastic straw would produce more GHGs than a paper straw. Ally Hirschlag wrote an article for the BBC, which was published on November 5, 2023, titled "Plastic or paper? The truth about drinking straws," where she discussed research completed by researchers at the University of Antwerp, Belgium. This research project discovered more hazardous chemicals in paper and bamboo straws than in plastic straws, and these chemicals can stay in the environment for a very long time. The straws might be more biodegradable, but what they release when they decompose could be more harmful than a plastic straw just sitting there for 200 years. Again, a plastic straw sitting in a landfill for 200 years, or a turtle getting it lodged up its nose and dying, or a paper straw decomposing and releasing chemicals into the surrounding area are all bad things, but they do not directly affect climate change or global warming at all.

When I am referring to waste in this section, I am talking about a by-product of a process which has no apparent use, excess material/products, or something which does not serve its intended purpose anymore. Just think about what you have disposed of today, maybe tea bags or coffee grounds, paper cups, banana peel/apple core, or maybe a drink container; all of those are recyclable by the way, and I hope you did your best to not just throw them in the garbage can. Waste can also include human waste when you go to the bathroom, and yes, methane is generated from that, which is a greenhouse gas, and for that reason and many others, it needs to be managed properly. Food waste is also a major issue today, and as mentioned before, depending on the food, a lot of GHGs are generated to produce it, large areas of forest or wetlands might have been destroyed to grow it, and it produces large amounts of methane as it decomposes in the landfill. Household hazardous waste like electronic waste (e-waste), cleaners, batteries, paints/solvents, drain cleaners, etc., even old smoke detectors and thermostats, potentially contain radioactive waste and mercury. Waste must be managed properly as it can cause harmful chemicals and pollutants to contaminate the environment and kill the surrounding wildlife. One could argue that preventing the negative effects of pollution damaging the environment by not properly managing waste would have a direct effect on climate change, as it would be destroying the natural carbon sinks, but for this book, we will keep climate change and waste management separate. Food waste is mainly generated from the food that grocery stores don't sell, goes bad, and has to be thrown out. There is also a lot of food waste from the producers, as many stores will only accept the most aesthetically pleasing products, because if the products don't look ideal, their customers will not buy them and it will be a big waste of their money. We need to change our mentality and start remembering that if a cucumber isn't perfectly straight, an egg's shell is a different color, a tomato isn't perfectly shaped, it will still taste the same and be just as good for you as the aesthetically pleasing items. If we can start showing the grocery stores that we will accept the more naturally looking items, there will be less waste from the producers, they will have more money in their pockets, and we will be generating less waste as a society. Another main source of food waste, other than from our homes is from restaurants, just think of all the times you don't finish your plate and they just throw it out; what I like to do is get it to go and give it to a homeless person on the street if I don't want to take it home.

A major issue with waste is that sometimes it is just excess product, which means too much was generated or purchased, or less was used than anticipated. An example of this is the groceries that we buy and don't end up needing; they end up spoiling, and we must dispose of them. This is not just a waste of money for us, but it is harmful to the environment as the rotten and decomposing food waste now has to be disposed of and properly managed. Waste management also includes recycling, which means certain waste streams can be processed and reused to produce other products. Recycling also prevents the need for the extraction and production of new raw materials, which would prevent further destruction of carbon sinks, prevent the generation of additional GHGs and have an overall positive impact on the climate change issue. There are parts within the waste management side that directly affect global warming and climate change.

Daniil Filipenco wrote an article titled, "World Waste: Statistics by country and short facts," published on March 7, 2023, and he wrote:

Annually, our planet becomes home to about 2.01 billion tons of municipal solid waste (MSW), and that figure is anticipated to reach 3.40 billion tons in about 30 years (a 70% increase).

According to the World Bank, high-income nations will show a daily per capita waste increase of 19%, while the same figure for low- and middle-income nations will be 40% or more.

MSW is garbage composed of all the items that people use on a daily basis, including food packaging, clothes, bottles, leftover food, papers, electronics, and batteries.

Despite all the recycling technologies in use, less than 20% of waste is recycled annually, with the remaining 80% becoming part of landfill sites.

The article also directs us to the Global Waste Index for 2022, which shows how certain countries rank for waste produced per person and how it was disposed of; landfilled, recycled, or incinerated. The top three countries listed were South Korea, Denmark, and Germany, respectively, while the bottom three were Chile 36nd, Latvia 37th, and Turkey was 38th; the UK came in at 18th, Australia at 19th, the USA was 25th, and Canada was 31st. On average, in South Korea, each person was recycling 60.8% of the waste they generated, while in Turkey, each person was only recycling 11%.

"The World Counts" has the following interesting facts to highlight some of the dark side of the global waste management industry:

Types of waste

Waste in many different forms is dumped on the planet. Wikipedia lists over 50 different types of waste. We go through a few of the worst ones below.

Electronic waste

The amount of electronic waste (or e-waste) has been exploding in recent years, along with technological developments. Roughly 50 million tons of e-waste are produced each year, and most of it is shipped to / dumped in poor countries in Asia and Africa.

It's toxic

E-waste is highly toxic if not treated properly. And sadly, that is often the case when dumped in poor countries. Heavy metals and toxic chemicals pollute surrounding communities and poison the local population, including children, and not least the people working at the dumpsites, often with no protection.

E-waste dump of the world

Guiyu, China, may be the world's largest e-waste dump with 700 tons of trashed electronics per year.

Really dirty stuff

Heavy metals and chemicals in e-waste include:

- Arsenic
- Lead
- Cadmium
- Chromium
- Flame retardants
- Vinyl chloride
- Dioxins
- Antimony
- PCB, PCDD, PCDF, DDT… WTF!

These heavy metals and chemicals are known to affect (not in a good way, in case you doubted):

- The central nervous system
- The reproductive and endocrine system

- The immune system
- Our blood, skeleton, digestive, and respiratory systems… basically everything in the human body.

Hazardous chemical waste

Industrial chemical waste is often very toxic and difficult and costly to dispose of. Some choose to "solve" this problem by exporting the toxic waste to countries with little or no control, where it is simply dumped without any treatment or safety measures.

In one example, 4,000 tons of toxic waste was shipped to Nigeria and simply dumped. The waste included 150 tons of highly toxic polychlorinated biphenyls, or PCBs, that are known to cause hormone deficiencies and reduced cognitive development in children.

The chemicals leaked into the local river and caused the death of nineteen villagers.

The dirty business of shipbreaking

Another type of waste often dumped in less developed countries is old discarded ships. It's much cheaper to send these ships to Asia to be broken down. 70% of ships are simply run ashore to be dismantled - mostly to China, Bangladesh, and India.

Shipbreaking is a very dirty business. Old ships often include toxic materials that contain, for example, lead, zinc chromates, mercury, arsenic, asbestos…

The workers often dismantle the ships without any protection.

Some materials are recycled. But not all. The rest is simply left on the beach:

"Stockpiled in Bangladesh, for example, are 79,000 tons of asbestos, 240,000 tons of PCBs, and 210,000 tons of ozone-depleting substances."

- World Bank

Global waste trade

Toxic and hazardous wastes are often sold by developed countries to poorer developing countries. This practice is known as the global waste trade. Wealthy nations are basically exporting their problem to poor countries, mainly in Africa and Asia.

Toxic colonialism

The practice of using underdeveloped states as a cheap way for richer countries to get rid of toxic waste is sometimes referred to as toxic colonialism.

Reverse logic

The rich countries exporting their waste have the technology and resources to process the waste safely. While the countries importing the waste most often do not have such resources, meaning that toxic waste is handled by people with no protection or knowledge about how to handle it.

Example of the toxic waste trade

In one case, 4,000 tons of toxic waste were shipped to Nigeria, falsely labelled as "fertilizers". The waste included 150 tons of highly toxic polychlorinated biphenyls, or PCBs, that are known to cause hormone deficiencies and reduced cognitive development in children.

The toxic chemicals leaked into the local river, causing the death of 19 people. The company made $4.3 million from the dirty trade.

Consequences of dumping waste in poor countries

Our waste is dumped in local communities thousands of kilometers away. The global waste trade has created enormous dumpsites in countries importing waste from developed countries of the West.

Affecting 64 million people

The dumpsites created by the global waste trade have at least one thing in common: They pose a serious threat to human health and the environment in the communities where they are located.

About 64 million people are directly affected by dumpsites created by the global waste trade.

World waste dumpsites

Some of the biggest dumpsites in the world:

Agbogbloshie e-waste dump, Accra, Ghana

- Receives around 192,000 tonnes of e-waste annually
- Pollutes soil, air, and water and causes serious health threats for the 10,000 people making a living from sorting and recycling.

The Bantar Gebang dump in Bekasi, Indonesia

- Total amount of waste up to 40 million tonnes
- Another 230,000 tons of household waste is added every year.

Jam Chakro dumpsite, Pakistan.

- 202 hectares and 5,000 people
- Affects the life and health of 5 million people
- Waste and the consumer society

The negative effects of consumerism include the generation of enormous amounts of waste, depletion of natural resources, and pollution of the Earth. If everyone lived like Western consumers, we would need up to 5 planets to support us and absorb our waste. And the number of consumers is growing by the second.

Municipal Waste Management

Municipal Solid Waste (the stuff you throw away at home) management is both your and the municipality's responsibility. Your local government should be using your tax dollars to set up proper waste management facilities for you to ensure the generic waste which you generate at home (garbage, sewage, and recyclables) has a safe place to go. If you live in an area which doesn't have proper waste management systems in place, I am sure you understand the importance of them, while if you are lucky enough to live somewhere that has proper waste management system, I hope you appreciated having your garbage taken away by someone else and stored in a safe location, and that you are able to go to the bathroom in your house and, with a flush of your toilet, the waste is not your problem anymore.

There is a lot of planning needed for creating and maintaining landfills as well as transfer stations. Please do not confuse the two or call either of them a dump or dumpsite. A landfill is the final home for garbage, where it is buried for the rest of time. If constructed properly, it has a liner, a contaminated water collection system, the waste is compacted into cells that are capped off with soil, has a methane collection system, as well as monitors and wells to ensure the landfill is not harming the environment. A transfer station is where the waste is collected and then transported to a landfill. It could have similar controls in place as a landfill. A dump is just an area where someone decided to pile garbage; it does not have any controls in place to prevent the garbage from harming the environment.

A properly designed landfill starts off in a specifically chosen area, which should not be near any environmentally sensitive areas. The first step to constructing a landfill is to dig a big hole. Depending on the ground composition, a thick rubber mat (liner) would be laid down on the bottom of the pit, and all the seams need to be sealed up to prevent any contaminated liquids (leachate) from being released into the environment. Once that is complete, the landfill should have a collection system installed to help collect the contaminated liquid, which might be pooling at the bottom of the pit. Material like sand or gravel will be placed over the leachate collection system to enable the leachate to make its way down into the leachate collection system. Fencing should be put up around the landfill to help catch any of the waste that could be blown away, as well as to keep out wildlife. Once that is set up, you have the basic controls in place to start accepting waste. Methane collection systems, air quality, and water quality monitoring systems should also be utilized to ensure the controls in place at the landfill are working to prevent harm to the surrounding environment. Once the landfill is set up and the waste starts coming in for its final resting place, it must be put into small sections called cells, and the waste must be compacted as much as possible to ensure the landfill will fit as much waste as possible.

A landfill only has a certain amount of space on the ground it can utilize (its "footprint"), so they must make the most out of the space they have to work with by building it as high as possible. It is very important that we divert as much waste from landfills as possible to give the landfill the ability to accept waste that really has nowhere else to go. When the cells are full or the landfill is maxed out, a layer of dirt is placed over it and compacted to seal it off or cap it. If you were to drive past an old landfill that was properly closed, it would probably just look like a grassy hill in the middle of nowhere, and you would not be able to tell that in that hill are decades of people's garbage. Issues with landfills include how gases like methane can build up from the decomposing organic material in the landfill; methane is not only a strong greenhouse gas, but it is also highly flammable. There have been cases where out-of-control fires and explosions occurred at some poorly managed landfills due to the buildup and ignition of methane. Properly constructed landfills collect the methane and either just burn it on site or use it to heat local buildings and homes.

Transfer stations are where most people go to drop off their garbage and recycling. They are basically where the waste materials are stockpiled until there is enough to load up a big truck and transport the waste to the landfill for final disposal. Properly constructed transfer stations should also have liners and leachate collection systems, as well as monitoring equipment to ensure the activities there are not negatively affecting the surrounding areas.

Technology is advancing regarding municipal solid waste disposal. Waste-to-energy facilities burn the waste, use the heat to heat up water and turn it into steam, the steam then spins a turbine and creates electricity which is then pumped straight onto the power grid and used as clean energy; this technology not only helps manage the issue regarding the never-ending generation of waste, but it also creates clean energy, a win-win. It is considered clean energy since there is very little smoke created as the garbage is burned at such a high temperature, and any smoke that is created is filtered, so the facility only discharges clean air. Besides dirty air filters, the main waste it produces is a small amount of ash left over from incinerating the waste, which is disposed of at a special landfill as it is considered a hazardous waste due to all the contaminants which didn't burn, like heavy metals. The residual heat and hot water left over can then be used to heat local homes and buildings. Municipal solid waste is constantly being generated all over the world and will always need a place to go, and with populations increasing, available land is shrinking. Incinerating it (burning it at a very high temperature) is the best solution, especially if we can create clean electricity and heat homes from it.

As I previously mentioned, proper waste management is your responsibility as well. To ensure you are not harming the environment, it is important to separate the recycling and household hazardous waste from the general garbage materials for many reasons, including:

1. Household hazardous waste could be dangerous for the workers processing the waste or the environment if not processed properly. This would include lithium-ion batteries (which are becoming very common), household cleaners and chemicals, and flammable liquids (including some paints).
2. Recycling helps prevent landfills from filling up and not being able to accept any more waste, as well as generates material which can be used for the production of other products, and reduces the amount of raw materials required from mining or manufacturing

3. Separating food waste also helps prevent the landfills from reaching their maximum capacity, as well as helps generate compost for plants.

It is also important to understand the limitations of your local recycling facilities, as there are different processes to recycle the different waste streams, and your local recycling facility might not be able to recycle them all and will just have to throw them away. This local knowledge would come in handy when you are planning your shopping. I don't want you to go and buy a bunch of the tetra pack juice boxes for the kid's lunches, thinking they are recyclable, and having the recycling depot in your area just throw them out because they don't have the technology to separate all the layers of the container.

One thing you can do if you are interested in reducing the amount of waste you generate is to conduct a waste audit on yourself. It is very simple; all you must do is let your waste container fill up as it normally does, and then once it is full, find an area where it is easy to dump it out and clean up after. Once you find your area, dump out your garbage and spread it around; you can also just leave the garbage in the garbage container and go through it that way if you like, but it is just easier to spread it out. Once you have the waste spread out, go through it, and make notes on what it consists of; once that is complete, put everything back in the garbage container. You are done with the hard part! It could get messy, so rubber gloves are recommended, and make sure that you disinfect the area the garbage was in once it is back in the disposal bin to ensure a safe living environment. Once you are done, review your notes and try to figure out ways to reduce the amount of waste you are disposing of. If your garbage bin is full of food waste, try to setup some sort of composting system, plan your shopping / cooking / food storage better etc. If it is full of recyclable cans and bottles, try to separate and recycle them better, if it is full of plastic bags, wrappers etc., maybe try to bring a reusable bag with you or try to buy something similar but with less packaging.

Unlike residential waste, human waste is something we cannot avoid We can buy less things, be mindful about the wrapping and packaging of what we buy are wrapped in, plan our shopping / cooking / food storage better so we do not produce any food waste, and even if we setup our lives in a way that gives our buddy (the weekly waste collector) nothing to do, we will still have to deal with human generated waste. The safe management of human waste is a very important part of a healthy society, but is often underappreciated by people who have access to it.

Depending on where you are in the world, the waste your body creates is dealt with differently. If you have a working sewage transportation system, you are luckier than most. According to a publication, part of the Biological Conservation in an article titled "Sewage pollution, declining ecosystem health, and cross-sector collaboration" by Stephanie L Wear, Vicenc Acuña, Rob McDonald, and Carme Font published in March 2021:

The lack of adequate sanitation for humans is a staggering problem. At least 4.5 billion people live without or have inadequate sanitation. In most of Sub-Saharan Africa, and on many tropical islands, more than 50% of the population lacks any form of sanitation. Even in countries with high coverage of modern sanitation facilities, poor design and lack of commitment to maintain infrastructure lead to massive pollution via discharge into surface and ground waters. Annually, more than 1.2 trillion gallons of untreated sewage, stormwater, and industrial waste flow into United States rivers because of overburdened treatment systems (Ahuja, 2014). Annually, 27 billion gallons of untreated sewage flow into New York Harbor alone (Riverkeepers, 2011).

Contamination of surface waters by fecal viruses is one indicator and dimension of the pollution problem. Accompanying feces and fecal viruses in sewage discharge are a cocktail of pollutants, including endocrine disruptors, heavy metals, pharmaceuticals, and pathogens (Wear and Thurber, 2015). These chemicals, like fecal viruses, are good indicators of sewage contamination. In Puget Sound, Washington, U.S.A., for instance, 25 of more than 80 contaminants found in effluent from sewage treatment plants were detected in nearby waters (Meador et al., 2016).

Another example of global modeling of sewage contamination uses a contaminant of emerging concern (diclofenac) and is shown on Fig. 2B (based on Font et al., 2019, and Acuña et al., 2020). Diclofenac is a common nonsteroidal anti-inflammatory drug that is only 40% removed by wastewater treatment plants (Font et al., 2019), and diclofenac persists in rivers and streams for weeks to months once released. Mapping of diclofenac in freshwater systems (Fig. 2B; Font et al., 2019) shows that sewage contamination is most likely to occur in densely populated regions where many people lack sanitation (e.g., India). These data also show that even in countries with near-universal sanitation, like Germany, there can be high loading of diclofenac into aquatic systems. Importantly, there was marked convergence in the diclofenac map (2B) and the predicted values of fecal viruses (Fig. 2A, B) in predicting global hotspots in sewage contamination. Both approaches produced similar spatial results that showed areas of high contamination levels over large areas in developing and developed countries (India, China, UK) and in hot spots around cities in other highly developed countries (US, Germany).

Even if you are lucky enough to live somewhere that has an adequate sewage facility (wastewater treatment facility), the facilities vary drastically with how much they treat and remove before discharging it back into waterbodies (oceans, rivers etc.). The most basic form of wastewater treatment is when the sewage (wastewater) is just put through a mesh screen or bars to remove large debris that was mixed into it, then the sewage is discharged directly into the water bodies. The next level of wastewater treatment facility is called a primary treatment facility, which is when only the solid material in the wastewater is removed and the remaining untreated liquid is discharged into the water bodies. The next level of wastewater treatment facilities is called secondary treatment, since it has two treatment processes; the first is the same as the primary treatment facility, as it removes the solid waste from the wastewater, and then the second phase of the treatment process is treating the remaining wastewater and killing off certain bacteria before discharging it. The third level of wastewater treatment facilities has three phases and is called a Tertiary Treatment facility; the first phase of the process is the solid removal, the second phase is the treatment of the remaining liquid, and the third phase of the tertiary treatment is a second treatment of the remaining liquid waste to remove specific contaminants before releasing the water back into the waterbodies.

Each of these different types of facilities has a different price tag on it, and your municipality might only have a primary treatment facility because that is all they can afford, or think is required. Where the treatment facility is built will also play a role in what level of treatment your wastewater needs. For example, if you are on an island in the ocean where there are strong tides and lots of water movement in the ocean, you could probably get away with just a primary or even with a simple screening process since the amount of wastewater being produced from the smaller island population as well as the wastewater quickly being diluted and turned into fish food. Now, if you live inland, let's say close to a large lake which people swim in and use recreationally, and that lake is the only place to discharge the treated wastewater, the municipality will need at least a secondary or tertiary treatment facility to ensure the harmful bacteria is at levels low enough to not negatively affect human and environmental health of the people who will be in close contact and possibly swimming in the treated wastewater. There are beautiful tropical islands around the world, many of which you know of and have wanted to visit at some point which have issues regarding clean drinking water, so what they have to do is recycle their municipal wastewater and clean it to a point where it meets local tap water standards; that would require a have to be a tertiary wastewater treatment facility.

I bet that you have been there trying to figure out what happens to all the solid waste which is removed from the primary, secondary, and tertiary treatment processes. Well, my friend, I am pleased to introduce you to bio-solids. This might be a bit of an early introduction as the untreated solid waste removed from the municipal wastewater treatment process still has to go through a couple processes including chemical and physical treatments to remove excess water and ensure the bacteria levels in the remaining solid waste is down to levels which are safe and do not pose as a risk to human health. Once the solid material is treated, you are left with a solid, soil-like material that is high in nutrients (for plants). This is a bi-product which is constantly being created at any municipal wastewater treatment facility and is generally just sent to a landfill to be buried. People are trying to find a good use for it as it can be used as industrial fertilizer, fill for closing mines during the reclamation process, capping landfills or something similar. I would not recommend using it in your garden at home, but it should be safe for industrial applications. Going back to the beginning of this book, I don't know if you remember, but I completed a research paper which five people have read, titled, "Exploring the Effectiveness of Bio-solids from Metro Vancouver's Municipal Sewage Treatment Process in the Bio-Degradation of PAH Contamination in Marine Sediments" in which I explored how different additive to bio-solids could help the natural bacteria in the bio-solids break down the hydrocarbon contamination from the polycyclic aromatic hydrocarbons (PAHs) in marine sediments. The idea was that we could repurpose this continuous waste stream to clean hydrocarbon spills and contamination. If you want to read it, let me know, I will be happy to provide it to you. As I mentioned, I tried to get it published, but the scientific journals did not like my statistical analysis and didn't want to publish it. One thing I want to add in here is that I have heard from people who work at municipal wastewater treatment facilities that it is not recommended to flush the "flushable" wipes; even if they are advertised as flushable, they end up wreaking havoc on the treatment processes as they do not break down like toilet paper does.

Regarding the unknown unknowns which are now known knowns, one major issue which has become apparent is how all of the medications which people are consuming are ending up in the water bodies. Sometimes people flush their unused medications down the toilet, but the main issue is the excess medication, which is not being absorbed by the body and is removed from the body when they go to the bathroom. Municipal wastewater treatment facilities are not setup to remove these types of contaminants, as they would require different processes for different medications and would be far too expensive and complicated to install. One major problem related to this is elevated levels of estrogen found in water bodies around where treated wastewater is discharged, and it has a direct effect on the aquatic life. Kundan Samal, Saswat Mahapatra, and MD Hibzur Ali wrote a paper called "Pharmaceutical wastewater as Emerging Contaminants (EC):

Treatment technologies, impact on environment and human health" published on June 16, 2022. It mentions the following:

In recent times, water bodies have been affected by Emerging Contaminants (ECs), which have the potential to sneak into the ecosystem and cause adverse impact on human health and ecology. Pharmaceutical contaminant (PC) is one of the major worrying classes of ECs, which arise from pharmaceutical industries that are biologically active compounds used to prevent, cure, or treat diseases. Personal care products (PCPs) are mainly used to improve the quality of daily life, which include lotions, detergents, hair dyes, lipsticks, cosmetics, creams, bath soaps, dental care products, shampoos, toothpaste, Sunscreens, fragrances, and other household items.. Pharmaceuticals and Personal Care Products are considered the source of pharmaceutical contaminants in the environment. There has been a massive hike in the manufacturing and use of PPCPs over the last few decades. Though there has been a massive surge in production and use of PPCPs, the treatment of the PCs released from the PPCPs has not been able to keep pace with the production. When the effluent from PPCPs is discharged into the water stream, it leads to genotoxic, mutagenic, and ecotoxicological effects to plants, animals, and humans. Constant release of PCs to water bodies and their exposure may result in long-term (chronic) effects in aquatic plants and animals. Jukosky et al. found that vitellogenesis is induced in male *Oryzias latipes* (Japanese Medaka) by estrogen. The mortality rate of fish is also increased by high estrogenicity. Exposures to PCs may also lead to changes in inherited traits and the behavior of living beings. Conversion of male fish into female ones by the development of female traits due to the presence of estrogen in water bodies is one such example of a shift in inherited traits. Moreover, PCs in drinking water can cause harmful effects in newborns, the elderly, and people who are suffering from kidney or liver failure. The occurrence of estrogens in drinking water can also affect men fertility too. It can further increase the incidence of breast and testicular cancer.

The elevated levels of estrogen in the waterbodies are mainly from birth control pills, in case you were curious, but just think of all the other medications people take, which are ending up in the waterways, to which the aquatic life are being exposed. I don't want you to feel guilty for taking your medications; your health is very important, and you should do what your doctor recommends above all else.

Industrial / Hazardous Waste Management

Industrial waste management refers to the waste generated from an industrial operation, and does not include the air pollution and GHGs that it might be producing. Examples of hazardous industrial waste would be a flammable/toxic sludge on the bottom of a fuel storage tank at a refinery, flammable paints, acidic wastewater from metal plating facilities, or even naturally occurring radioactive material (NORM), which occurs naturally during some industrial operations; there is also non-hazardous industrial waste. There are processes in place to deal with all waste streams, be it solid, liquid, gas, or sludge, explosive, compressed gas, flammable liquids, flammable solids, oxidizers, toxic, radioactive, corrosive, or anything else; it will just cost the generator (owner) of the waste a lot of money.

Industrial waste can also have one or more hazardous characteristics; it could be flammable and corrosive, or explosive and an oxidizer. If some of the different waste streams are mixed, they can create entirely new hazards. For example, if a strong acid and an oxidizer accidentally become mixed, it can create a very hazardous environment. The safe packaging, labeling, transportation, and processing of these hazardous materials takes years to master, and one would need to have a basic understanding of chemistry to do it; it's not just about compliance with regulations, it's about ensuring the safety of the surrounding people. Mishandling hazardous chemicals and substances can be deadly, just look at what happened in Beirut, Lebanon, on August 4, 2020, or that train derailment in Ohio on February 3, 2023. Depending on the hazard which each waste stream has, they require different handling and processing; some can be incinerated, others have to be put back into the Earth down a deep hole and buried in concrete so it cannot come into contact with anything again; some can be sent to a special landfill with extra controls in place, and some just need to be blown up on site because it is too dangerous to move.

Unfortunately, there are many industrial operations around the world that do not care to properly deal with their waste and cause massive environmental disasters. The issue of dealing with their industrial waste properly can be very expensive for the waste generator. Unlike municipal waste disposal, which is basically free (paid for by taxes), the industrial waste producers need to pay every time they dispose of something, and it is generally not cheap. For these industrial operations to save money, they illegally dump their waste into the environment somewhere, potentially harming both the wildlife and human health in the areas. I don't know if you remember Erin Brockovich, but she is an American paralegal who helped build a case against a gas and electrical company regarding their industrial operations contaminating the groundwater, which made the people in a nearby town very sick; there was a movie made about the events, and Julia Roberts was amazing in it. Unfortunately, issues similar to what happened to the town that Erin Brockovich was fighting for happen all the time around the world, and I am sure it is worse in developing countries where people might not have the ability to stand up for themselves and get the help they need.

There are also accidents which occur, containments for hazardous waste can fail and release whatever it was storing back into the environment, causing massive issues. This rarely happens, but sometimes big dams that hold waste rock from mining operations (tailings) can break and release the contaminated sludge/wastewater downstream and cause major issues to the surrounding areas. A mining process called fracking, if done incorrectly, is known to contaminate groundwater, which can spread and contaminate the surrounding areas and potentially people's drinking water. Depending on where you are in the world and how strict the government regulations are, you and the environment can be perfectly safe with the tailings dams and fracking.

There are many contaminated sites around the world, and they have become contaminated from spills, leaks, improper storage of products or waste etc. Contaminated sites can consist of contaminated soil, groundwater, or both, and are the responsibility of the landowner. Unfortunately, many contaminated sites have been abandoned and are causing human and environmental health issues. In the US, in 1980, Congress created the "Comprehensive Environmental Response, Compensation and Liability Act (CERCLA)" in response to growing public concern over toxic waste dumps, including Love Canal and the Valley of the Drums in the 1970s. The act is for the following:

This law created a tax on the chemical and petroleum industries and provided broad Federal authority to respond directly to releases or threatened releases of hazardous substances that may endanger public health or the environment. Over five years, $1.6 billion was collected, and the tax went to a trust fund (superfund) for cleaning up abandoned or uncontrolled hazardous waste sites. The Comprehensive Environmental Response, Compensation, and Liability Act of 1980 (CERCLA):

- established prohibitions and requirements concerning closed and abandoned hazardous waste sites;
- provided for liability of persons responsible for releases of hazardous waste at these sites; and Established a trust fund to provide for cleanup when no responsible party could be identified.

The law authorizes two kinds of response actions:

- Short-term removals, where actions may be taken to address releases or threatened releases requiring prompt response.
- Long-term remedial response actions that permanently and significantly reduce the dangers associated with releases or threats of releases of hazardous substances that are serious, but not immediately life-threatening. These actions can be conducted only at sites listed on EPA's National Priorities List (NPL).

The following table shows the number of Federal and general sites for each status and milestone as of March 07, 2024:

Status	Non-Federal (General)	Federal	Total
Proposed NPL Sites	37	2	39
NPL Sites	1183	157	1340
Deleted NPL Sites	439	18	457

If you are interested in knowing what sites are approved or proposed for the EPA to fund its cleanup, you can find it here https://www.epa.gov/superfund/superfund-national-priorities-list-npl.

Manufacturing Waste Management

I separated manufacturing waste and industrial waste management to highlight the different waste streams when compared to industrial waste management, even though there are many similarities. Wastewater management is generally a similar waste stream that comes out of both industries, as well as certain waste materials, but the waste materials from the manufacturing industry tend to be able to be reused/repurposed more than the industrial waste and are generally much less hazardous. The clothing manufacturing industry is a major contributor of waste, as is the fashion industry as a whole. Martina Igini wrote another article for Earth.org, which was published on August 21st, 2023, titled "10 Concerning Fast Fashion Waste Statistics," and it mentions the following:

The extremely detrimental impact of fast fashion waste on the environment is no news. Besides being responsible for nearly 10% of global carbon emissions, the industry is also infamously known for the amount of resources it wastes and the millions of clothes ending up in landfills every day. Here are 10 highly concerning statistics about textile waste.

1. 92 million tonnes of textile waste are produced every year

- Of the 100 billion garments produced each year, 92 million tonnes end up in landfills. To put things in perspective, this means that the equivalent of a rubbish truck full of clothes ends up in landfill sites every second. If the trend continues, the amount of fast fashion waste is expected to soar up to 134 million tonnes a year by the end of the decade.

2. The apparel industry's global emissions will increase by 50% by 2030

- If a business-as-usual scenario prevails in the coming years – meaning that no action is taken to reduce fast fashion waste – the industry's global emissions will likely double by the end of the decade.

3. The average US consumer throwsaway 81.5lbs of clothes every year

i. In America alone, an estimated 11.3 million tons of textile waste – equivalent to 85% of all textiles – end up in landfills on a yearly basis. That's equivalent to approximately 81.5 pounds (37 kilograms) per person per year and around 2,150 pieces per second countrywide.

a. The number of times a garment is worn has declined by around 36% in 15 years.

i. The throwaway culture has worsened progressively over the years. At present, many items are worn only seven to ten times before being tossed. That's a decline of more than 35% in just 15 years.

b. The fashion industry is responsible for 20% of global wastewater.

i. Dyeing and finishing – the processes by which colour and other chemicals are applied to fabrics – are responsible for 3% of global CO2 emissions as well as over 20% of global water pollution. Along with yarn preparation and fibre production, these two processes have the highest impacts on resource depletion, due to the energy-intensive processes based on fossil fuel energy.

c. It takes 20,000 litres of water to produce one kilogram of cotton.

i. Besides being a huge source of water pollution, fast fashion also contributes to massive quantities of water being wasted every day. If this is difficult to picture, just think that about 2,700 litres of water are needed to make just one t-shirt, which would be enough for one person to drink for 900 days. Moreover, a single load of washing uses between 50 and 60 litres of water.

d. $500 billion is lost each year because of under-wearing and failure to recycle clothes.

i. The worst aspect of our reckless, throw-away culture is that most clothes being tossed each year are not recycled. Globally, just 12% of the material used for clothing ends up being recycled. Much of the problem comes down to the materials our clothes are made from and the inadequate technologies to recycle them. "The fabrics we drape over our bodies are complex combinations of fibres, fixtures, and accessories. They are made from problematic blends of natural yarns, man-made filaments, plastics, and metals."

e. Nearly 10% of microplastics dispersed in the ocean each year come from textiles.

i. Garments are a huge source of microplastics because so many are now made of nylon or polyester, both durable and cheap. Each wash and dry cycle, especially the latter, sheds microfilaments that move through our sewage systems and end up in waterways. It is estimated that half a million tons of these contaminants reach the ocean each year. That's the equivalent to the plastic pollution of more than 50 billion bottles.

f. 2.6 million tons of returned clothes ended up in landfills in 2020 in the US alone

i. Most of the items returned to retailers from consumers end up in the landfill. This is mainly because it costs more to the company to put them back in circulation than to get rid of them. Reverse logistics company Optoro also estimates that in the same year, 16 million tons of CO2 emissions were created by online returns in the US in 2020 – the equivalent to the emissions of 3.5 million cars on the road for a year.

1. Fast fashion brands are producing twice the amount of clothes today than in 2000

This dramatic increase in production has also caused an increase in both pre- and post-production textile waste. Due to the number of cutouts for the clothing, many materials get wasted as they cannot be used any further, with one study estimating that 15% of fabric used in garment manufacturing is wasted. Post-production, 60% of approximately 150 million garments produced globally in 2012 were discarded just a few years after production.

There are many organizations that are looking to help repurpose the waste materials and unwanted clothes to help keep them out of the landfill; one of these initiatives includes collecting the unwanted clothes and shipping them around the world to people who do not have the ability to get new clothes. I have been playing with the idea of starting a company called "Ugly Bags" for the past eight years. It takes the waste material from the textile industry, as well as unwanted clothes, and pieces them together to create reusable shopping bags to help replace the plastic bags which many companies are still using. The bags would not look pristine, but more like Frankenstein's monster, but they would be iconic, a statement piece, proving a point, and showing everyone that your priority is to save the environment. It would be noticeable from far away, and whoever would be using it and proudly carrying it around would be showing the world that they are environmental champions. Regarding the business model, I am sure you could charge companies to take their waste textile since they have to pay to dispose of it themselves, then cut and sew the random pieces of fabric together to make a bag (not super challenging), and then sell it to consumers or businesses; making money from obtaining your material as well as when you sell your product and at the same time helping the environment in multiple ways sounds like a mega win. I am sure that companies will buy into this and want their logos on it to show that they also care about sustainability and are environmentally conscious. My plan included sending one bag for every bag bought to countries that have issues with plastic bags.

During my travels, one country I went to had a serious issue with plastic bags and waste management in general. It was heartbreaking since the city I spent most of my time in was on the coast, had beautiful beaches and warm water, but as soon as you go into the ocean, there are plastic bags floating just beneath the surface like a school of jellyfish. I quickly understood why that was, as soon as I went grocery shopping, the quality of the plastic bags, which all the stores seem to use, was very poor, probably because the owner of the store didn't have or want to spend the money on good-quality plastic bags. The cashier had to double or even triple bag my groceries to make sure the bag didn't rip and ensure I didn't drop my groceries everywhere on the walk to my accommodations. I asked the cashier to just use a single bag, but with broken English, they made it clear that it had to be double or triple-bagged; I could tell that the cashier didn't feel good about using so many plastic bags, but they don't have any other options. The thing about areas like this is that they probably do not have a proper landfill or waste management systems in place, which explains why so many of these plastic bags just get picked up by the wind or rainfall and carried down into the oceans. By providing reusable bags to countries like this, we would really be making a big difference in the number of plastic bags ending up in the ocean. It is not the developed countries that are putting large amounts of plastic waste (like bags and straws) into the ocean since we have proper waste management systems in place, including properly constructed and maintained landfills. It is countries that lack the ability to set up proper waste management processes and facilities that are the main suppliers of ocean-bound waste. It is not entirely their fault as they don't have the resources to set something like that up and they probably require outside support, which is currently not really happening.

Another thing I would like to see happen is for an organization to set up a reusable bag collection system. I know that almost every time I go grocery shopping, I forget to bring my reusable shopping bags and must buy new reusable shopping bags. Doing this, I now have probably 40 reusable shopping bags, and I only require a maximum of 4-5 bags, which means I have about 35 extra reusable bags that I will never have a need for and they just sit around; I am sure I am not the only person who has this issue. I think it would be beneficial if there was an organization that collected everyone's extra reusable bags and sent them to countries that don't have access to reusable bags and still rely heavily on plastic bags. Maybe if I had a couple of special reusable bags like the "Ugly Bags" I wouldn't forget them all the time.

Circular Economy

The Government of Canada says the following about a circular economy on its website, and was last updated on December 23, 2022:

The circular economy is a different way of doing business.

The way our economies extract, use, and dispose of resources is putting pressure on our natural systems, communities, and public health. This is a linear economy—it moves in a straight line from resource extraction to waste disposal.

In a circular economy, nothing is waste. The circular economy retains and recovers as much value as possible from resources by reusing, repairing, refurbishing, remanufacturing, repurposing, or recycling products and materials. It's about using valuable resources wisely, thinking about waste as a resource instead of a cost, and finding innovative ways to better the environment and the economy.

Both businesses and individuals can save money and create opportunities in a circular economy. For example, businesses can launch product buy-back programs that recover usable materials, or create new materials from product waste—like turning pulp-and-paper-mill waste into renewable plastics or textiles.

As individuals, we can use circular economy principles to cut our waste and save money by reusing or repairing items, or by using sharing services like tool libraries or car sharing. This movement has been building momentum as governments, businesses, and individuals embrace more circular approaches.

The World Economic Forum published an article on December 8, 2020, titled, "Circular economy examples – How Ikea, Burger King, Adidas and more are investing in a circular economy" which mentions a few initiatives major organization have made in the past, and includes the following:

Recycling incentives: Thousand Fell

Thousand Fell is already making a name for itself as an environmentally-conscious manufacturer with shoes made from sustainable materials such as coconut husk and sugar cane, and even recycled plastic bottles. Now, in partnership with TerraCycle and UPS, the maker has launched a special recycling incentive. Customers can return old pairs of Thousand Fell shoes back to the manufacturer. Thousand Fell will then recycle the returned footwear and send customers $20 that can be used toward a new pair of shoes.

A big brand selling goods second-hand: IKEA

Visitors to the Swedish town of Eskilstuna, about 100km outside of the capital Stockholm, could visit a 1,000-year-old stone covered with Viking runes and pictures. They could also visit IKEA's first-ever second-hand store.

The shop will feature gently used IKEA furniture as part of its efforts to reach its 2030 climate targets. Head of sustainability at the Scandinavian furniture giant Jonas Carlehed told Reuters earlier this year that: "We are making a huge readjustment, maybe the biggest IKEA has ever made, and one of the keys to reaching [the company's 2030 climate targets] is to manage to help our customers prolong the life of their products." The company has also recently started a buy-back scheme for customers – it gives vouchers in exchange for the return of unwanted furniture and other items. That scheme has, however, been suspended in some locations because of ongoing pandemic-related restrictions.

Reusable fast food packaging : Burger King

Takeaway food is big business -- but the packaging for those meals poses a sustainability challenge. Global takeaway brand Burger King has unveiled a solution in the form of reusable packaging intended to reduce the amount of waste it generates. Customers in New York, Tokyo, and Portland, Oregon will soon be able to buy burgers and drinks in reusable packaging. The plan, one in place for next year, features a small deposit charged initially and then refunded when the customer returns with the boxes and cups, which are taken away for cleaning and processing via the zero-waste e-commerce system Loop.

Shoes you don't own: Adidas

Adidas is another circular economy example. This sportswear multinational has a range of footwear designed with recycling in mind. Its UltraBoost DNA Loop shoes are made from just one material – thermoplastic polyurethane (TPU). No glue is used in its manufacture; instead, it is assembled using high temperatures. On its website, Adidas describes the UltraBoost Loop as the shoes customers will never own, but will instead return once they are finished with them. "If the end can become the beginning, we can help keep products in play and waste out of landfill," the company says.

It's been a couple of years since that article came out, and it looks like Thousand Fell shoes are doing well, Ikea's website still talks about their "circular agenda" and their 2030 goals, but I can't find much on the success of Burger King's reusable food containers or Adidas' recyclable shoes.

With all of the changes regarding sustainability, reusing materials, circular economy, I can't stop thinking about how we are getting closer to Zoolander's "Derelicte" fashion trend; it doesn't seem like such a far-off idea anymore. There is still a lot of room for improvement, and we need a lot of creative minds working on this to find synergies between industries and ways to reduce and repurpose waste from manufacturing and industrial operations. We also need everyone (including you and me) to change the way we deal with the products we do not want to use anymore and do our best to keep them in use by fixing them or selling them instead of just sending them to a landfill to sit for the rest of time, or worse, end up dumping in the environment somewhere.

Dump sites or dumps, refer to areas where someone decided to improperly dispose of their unwanted things. You might see them at the end of dead-end roads or a short walk into a forested area, or off to the side of the road on a gravel or forest service road. I am guessing people end up throwing their unwanted items in these dump sites because they don't want to pay to dispose of them properly at the landfill or transfer station, or they didn't know how to dispose of them properly. Regardless of why, these dump sites can be hazardous to the local environment as the waste will react with rainwater and potentially create a harmful sludge containing heavy metals or other contaminates. These contaminates could be absorbed into the ground, poisoning the area or migrate from the surface or groundwater and end up in rivers and creeks, poisoning the wildlife it encounters on the way. Household hazardous waste is commonly found in these dump sites, along with other hazardous waste like asbestos-containing materials and debris with leachable lead-containing paint.

In Canada, we have an organization called "Product Care" which helps people safely dispose of certain types of their household hazardous waste for free. How it works is the manufacturers and producers of the household hazardous products pay a small fee to "Product Care" which is then used to pay for the disposal of the waste material if there is any. It is always good to contact your local landfill, transfer station, or recycling depot to just ask what they accept and what can be recycled; they should be able to guide you in the right direction for proper disposal of specific items in your area.

Ocean Waste Management

A lot of the solid waste we create on land has made its way into the oceans, and a lot of that is plastic waste. There are also contaminates that have made their way into the ocean through rivers, creeks, from people dumping and discharging their contaminated wastewater into the ocean from industrial operations, manufacturing operations, agriculture/aquaculture operations, as well as other industries. It is not okay to pollute our oceans with contaminants, but for the sake of the book, I am going to focus on the solid waste that is in the ocean; I will discuss where it came from, what it is, how it got there, and why we need to clean it up. I will also discuss current operations underway and a concept I have been working on, which could dramatically aid the efforts to not only remove the solid waste currently in the oceans but also prevent the oceans from getting to that point again.

The Ocean Conservancy organization states the following:

Every year, 11 million metric tons of plastic enter our ocean on top of the estimated 200 million metric tons that currently circulate our marine environments. Whether by errant plastic bags or plastic straws winding their way into gutters or large amounts of mismanaged plastic waste streaming from rapidly growing economies, that's like dumping one New York City garbage truck full of plastic into the ocean every minute of every day for an entire year! And that much plastic is bound to have an impact on ocean ecosystems.

An organization called AquaBlu put together the following information:

According to research from #PlasticFreeJuly and packaging retailer RAJA UK, India topped the list as the biggest polluter worldwide with a staggering 126.5 million kg each year that ends up in the ocean. China and Indonesia also topped the list as the biggest plastic polluters in our oceans.

Countries responsible for the biggest plastic pollution in our oceans in 2022:

1. India (126.5 million kg).

2. China (70.7 million kg).

3. Indonesia (53.3 million kg).

4. Brazil (38 million kg).

5. Thailand (22.8 million kg).

6. Mexico (3.5 million kg).

7. Egypt (2.5 million kg).

8. United States (2.4 million kg).

9. Japan (1.84 million kg).

10. United Kingdom (0,703 million kg).

Looking at this list, three things that immediately pop out to me are that the top two are countries with the largest populations in the world, three of the top five producers are developing countries, and the other two are classified as newly industrialized. I am guessing that most of these countries do not have proper waste management systems in place, including collection, transportation, and disposal, with such densely populated cities and no waste collection service. This list didn't surprise me. It seems to me that the only way for people who live in these areas to dispose of the garbage and have it taken away from their living area is to put it in a river and have the river carry it away instead of having it pile up in their homes and in their streets; I would probably do the same thing if I were living in that situation. It is not the people's fault that they do not have the processes and infrastructure to deal with their waste.

Now let's talk about why we need to clean up the plastic waste from the oceans. Aquatic life, as well as marine birds, see the plastic and think it is edible, and so they eat it. When they eat the plastic, it stays in their stomach since they cannot digest it. The plastic accumulates in their stomachs, and over time it eventually prevents the wildlife from eating actual food with the nutrients to keep it alive, and they end up dying of starvation or living a reduced quality of life. There are small and tiny pieces of plastic (microplastics) that are generated from plastic waste like pop bottles and plastic bags, which break down into smaller and smaller pieces over time in the oceans. As mentioned earlier, our buddies the phytoplankton eat the microplastics, which can then kill them, which can decrease the ocean's ability to absorb CO2. Larger zooplankton also eat the microplastics in the ocean, as well as the phytoplankton, which probably already ate the microplastics, so they would be getting a double dose of plastic in their diet and most likely end up dying as well. Zooplankton are consumed by larger fish, and humans can potentially consume those fish. Can you see how plastic can make its way up the food chain and how that affects us The microplastics harm fish in another way, since fish get oxygen from moving oxygenated water through their gills, when the fish try to draw water over their gills which contains microplastics, the microplastics can end up becoming trapped in their gills and prevent them from obtaining the oxygen they need to survive, essentially choking them. Plastics are being observed in most of the aquatic wildlife and sea birds all over the world, and it seems to be getting worse. The plastic in the ocean will never be removed naturally; it will just continue to break down into smaller pieces and contaminate all the ocean life and sea birds as well as make its way up the food chain to humans. It is up to us to prevent further waste from entering the oceans and to clean up what is there, not just to save aquatic life, but to help improve the ocean's ability to process CO2.

Due to the ocean's currents, five large garbage patches have formed around the oceans. The NOAA says the following about these garbage patches:

Garbage patches are large areas of the ocean where litter, fishing gear, and other debris - known as marine debris - collects. Rotating ocean currents form them called "gyres." You can think of them as big whirlpools that pull objects in. The gyres pull debris into one location, often the gyre's center, forming "patches."

There are five gyres in the ocean. One in the Indian Ocean, two in the Atlantic Ocean, and two in the Pacific Ocean. Garbage patches of varying sizes are in each gyre.

The most famous of these patches is often called the "Great Pacific Garbage Patch." It is in the North Pacific Gyre (between Hawaii and California). "Patch" is a misleading nickname, causing many to believe that these are islands of trash. Instead, the debris is spread across the surface of the water and from the surface all the way to the ocean floor. The debris ranges in size from large abandoned fishing nets to tiny microplastics, which are plastic pieces smaller than 5mm in size. This makes it possible to sail through some areas of the Great Pacific Garbage Patch and see very little to no debris.

Ocean Cleanup

The key to creating clean oceans requires a couple of different things, the first is we need to prevent any more waste from entering the ocean; that can be done by helping to implement proper waste management processes in the highest oceanic waste producing countries or setting up waste collection processes at the mouth of the rivers which feed into the oceans. The second is physically removing as much of the plastic as we can from the oceans, starting with cleaning up as much of the garbage patches as possible. As previously mentioned, the equivalent of a dump truck full of garbage is being dumped into the oceans every minute. If we do not stop the waste from entering the oceans, we need to remove one garbage truck worth of waste per minute just to maintain the amount of solid waste we have in the oceans currently; we will need to remove more than one garbage truck per minute to start making a dent in it. We have the processes, understanding and technologies to deal with solid waste management on land. We just need to help the major waste contributing countries set up proper waste management processes to prevent them from having to dispose of their waste in the rivers and oceans. This will meet the first objective. The second objective of cleaning up the oceans will be much more challenging since the oceans are so large and there are already approximately 200 million metric tons of waste to remove.

"The Ocean Cleanup" is the largest and most successful group currently working on cleaning up the oceans. They are addressing the issue of preventing further garbage from entering the oceans by collecting waste at the mouths of rivers which come from the major oceanic plastic waste producers. The Ocean Cleanup is also actively collecting and disposing of garbage from the open ocean. Their goal is to remove 90% of the floating plastics. Other organizations that continue to come up during my research are Blue Ocean Project, the International Coastal Cleanup, Surfrider Foundation, Oceana, Green Peace, and the Environmental Defense Fund. Many of these organizations fight for policy and law changes to help save the oceans or focus on beach and river cleanups to help prevent the garbage from entering the oceans.

The ocean waste cleanup technology I have been developing over the past few years draws on my experience working in the waste management industry. I believe that if we incorporate some of the successful ideas currently utilized in waste management processes and landfill design used on land, and incorporate it into processing waste on the water, we can really start making a dent in the ocean cleanup. I have been picturing an old cargo ship, restored and improved for the sole purpose of ocean waste management. Like the systems currently being used by "The Ocean Cleanup", there would be two smaller vessels that help funnel the waste into the main vessel, and like The Ocean Cleanup's setup, the refit cargo ship will have a conveyor belt which will feed the floating waste into its hull. The main difference between the idea I have been working on and what is currently being used is that the cargo ship would process the waste on the water and not have to return to land or transfer the waste to another vessel. As we know, the most effective way to store waste is to compact it and make it dense, and the best way to compact something is to shred it into small pieces to help it squish together. The design I have been working on is, once inside, the waste is transferred to a shredder to turn it into small pieces, then the pieces can be compressed and bundled into large car-sized garbage bricks. This would significantly increase the amount of waste a ship can carry as it reduces it's volume. The next step in the process is the game-changer. Remember the waste-to-energy facilities we were discussing earlier? The design I have been working on then takes the bricks of plastic and feeds them into the waste-to-energy process built into the vessel, and it could be a mobile waste-to-energy facility. The energy created could power itself, creating a hybrid, mobile waste-to-energy vessel. The vessels could be sent to different garbage patches around the world to help dispose of the waste; they could be parked at the mouth of the major waste transporting rivers and help intercept and process the waste before it enters the oceans. For a business idea, these mobile, hybrid waste-to-energy vessels could dock at major cities, incinerate their waste for a fee, and sell them back energy; sounds like a good business plan to me, which would work great for coastal developing countries. Regarding the oceanic waste, I don't think it is worth collecting all the garbage, sorting through it, separating what is recyclable and what is waste, since we are currently in such a state of emergency with this. Sorting the waste would be a better idea for the beach and river cleanups, but we need to get the garbage out of the oceans asap.

Solutions for Waste Management

By now, I hope you can see the differences between extreme climate change and the issues with waste management. There are some overlapping areas regarding the two, but I believe they are fairly independent of each other and each requires its own solutions. Human waste is always going to be something we will have to deal with, and if we can come up with a permanent use for biosolids, we would be in a much better situation. Waste is generated mainly from our purchases, many of which are not necessary. Just think about it, almost every time you go out and buy something and bring it home, you will be bringing waste into your house. To reiterate, an easy way to overcome this, save you some money and also help reduce the amount of GHGs your actions produce is to buy used items instead of new; there is no packaging to deal with, you are keeping something out of the landfill, and you are reducing the demand for manufacturers to produce more items, which means less GHGs will be generated. I understand that properly packaging items is important for transporting and storing them to help protect them, but if we could figure out a better way to package merchandise, we would be doing ourselves a favor. If you want to do your part to help reduce the amount of waste you produce, just be more mindful of what you purchase, and hopefully, manufacturers and suppliers will see the trend of people choosing products with more environmentally friendly packaging and start becoming more innovative and creating more sustainable packaging methods.

Food waste is another major waste stream that needs to be dealt with. Food waste is generated from expired food from grocery stores, discarded leftovers, and food from restaurants, and food that was thrown away in your home because it went bad, people just not finishing the food on their plate, or not wanting to save the extra food for later. Planning your shopping and food storage better will not only save you money, but it will also save the environment. Keeping leftovers or even sharing food with your friends, family, or even neighbors could help divert food waste from the landfills. Freezing food or just buying less are solutions as well. Grocery stores can do better with disposing of their rotting food, either sell it at a discounted price or for free just before it goes bad to try and give it another chance of being used. It is also our fault that grocery stores have so much food waste. I am sure that they buy extra because they don't want customers coming in and seeing empty shelves, and go to one of their competitor's instead next time. If we developed a mentality that it is good for the stores to have empty shelves since they won't have to dispose of so much waste, and we just buy something else that they have in stock, grocery stores wouldn't have to buy so much excess food and have so much to dispose of. Planning your shopping better to ensure you use all of the food you buy and prevent food waste will not only save you money but also save the environment.

If you didn't before, I hope you now have a stronger appreciation of the municipal waste management operations, especially for the hard-working men, women, and anyone in-between who help collect and process your waste to ensure you live in a clean environment. I hope you can appreciate the importance of having a properly designed and maintained landfill, and how unmonitored "dump sites" can lead to environmental issues, including toxic runoff into surrounding streams and rivers, and harm the local wildlife. Having properly set up waste management processes and infrastructure is fundamental to a healthy society. Imagine a world where your garbage was not collected and you had to deal with your own human waste, while living in tightly packed cities where you wouldn't be able to tell your garbage pile on the street from your neighbor's garbage pile. You would be fighting off rodents, insects, and bacteria every step of the way. Imagine a world where all the hazardous waste produced from industrial and manufacturing operations was left unchecked, and they dumped their environmentally hazardous waste into water bodies, forests, farmland, or wetlands; there would not be much of a healthy planet left. We can cut back on the amount of waste we produce through better planning and changing our buying habits.

95% of the waste that ends up in the oceans is generated by developing or newly industrialized countries, and it is apparent that they are struggling with setting up proper waste management processes. For regions with high, dense populations that don't have a lot of available land, utilizing waste-to-energy facilities would greatly benefit. They are expensive to build and maintain, but seeing as it would also generate clean power, it would be a win-win for both waste management and extreme climate change regarding clean energy production. By helping to keep the waste out of the oceans, we are giving the aquatic life and sea birds a better chance at thriving, helping the populations of phytoplankton to grow, and helping to reduce the amount of CO_2 in the ocean. Again, another win-win for the waste management and global warming issues.

Homework

A lot of what was discussed here has to do with major projects that will require a lot of research and funding to complete, and that might not include you. One thing you could do if you are interested in helping is to hold fundraising events, or create something like a "Go Fund Me" account to raise money to fund some of these projects or others that interest you. If you are able to financially support some of these initiatives, feel free to reach out to me, and we can figure something out; I can let you know which organizations or projects will make the most from your donation or guide you to an organization that is working on something you are passionate about. If you are just interested in how you can make some major or minor changes in your lifestyle to help do your part to get the planet back to where it should be, I will provide you with some ways you can do it. The great thing about many of the ways you can start doing to help the environment is that they also save you money, so why wouldn't you do it?

The obvious one, which I am sure you have heard people tell you since you were a kid, is to save electricity. With the world heating up, the increasing droughts and less precipitation, the hydroelectric generators might be running into some issues in the near future. If we don't start leaning more on alternative clean energy like nuclear or solar, we would need to rely mainly on energy-producing facilities that burn fossil fuels, which would make our situation much worse. Depending on where you are and your living conditions, you can upgrade the insulation and/or windows in your home, change your lights to LED, buy energy efficient appliances, unplug unused appliances, turn the lights off when not required, turn the heat down and wear a sweater or cozy up with a blanket, or turn down the A/C and use ice packs or cold damp cloths. What I have done in the past is put a shirt in water, make it soaking wet, then stick it in the freezer for a bit; once it is frozen, I would put it on and it would be the most refreshing experience. If you have 2-3 shirts cycling between being worn, frozen, and freezing, you should have enough there to keep you at a comfortable temperature during hot times. There is not one answer that will solve everyone's energy over-consumption since everyone has a different setup, but just think about what you use that requires energy and figure out a way to use less of it.

Reducing the amount of carbon-based fuel usage, be it gasoline or diesel. See what you spend most of your time driving for and figure out ways to reduce it. If it is to work, and you can work from home, maybe bring up to your manager how you would like to reduce your carbon footprint and would like to work from home. If you don't want to do that, just gift them this book, tell them it was life-changing and the greatest book since Silent Spring, and hopefully they will come to the idea on their own. I enjoy taking public transit because of all the interesting people and things I see on it, but I am also a 6'4" 260 lbs. male and don't have much to worry about when it comes to worrying about my safety, and I understand why many people aren't comfortable with it. I am sure it can be very unreliable and doesn't take you anywhere close to where you need to go, or it takes too long to get to your destination. A couple of other obvious changes you can make are to bike and walk more, but again, I understand the safety and time aspects of it, and with all the things we have to do in the day, it just makes sense to drive, but just try to cut back where possible with even something like carpooling. With the gas prices the way they are now and how they are trending, this one would definitely help you keep some of that hard-earned money in your account. You can also upgrade your vehicle to a hybrid or electric car, but just keep in mind that the electricity you are using might be from a dirty power source, and it could be doing more harm than good.

You can also investigate changing your vehicle so it can use alternative fuels.

Being kind to the environment will make a huge difference. I don't think you are out there cutting down trees for fun, pouring chemicals into water bodies, or just dumping your garbage in places you shouldn't, but if you spot people dumping in places they shouldn't be or if you come across some garbage that shouldn't be there, if it is safe to do so, clean it up, if not, report it to your municipality. Save water, try not to leave the faucet running, take showers instead of baths, and take quicker showers. Try to get creative and capture the water you use when you first turn the faucet on and are waiting for it to reach the temperature you are looking for, use it for watering any indoor plants you have, washing dishes, save it for next time you use the toilet and use it to fill up the tank of your toilet (you can turn off the water which fills up the tank with the valve which goes into the wall behind your toilet). Another thing you can do is if you have a dishwasher, only wash full loads or use the half load option if you have it to help make the water use more efficient. When washing clothes in a washing machine, try to use cold water and adjust the load size to what is appropriate. Again, all of these are things that will save you money and the environment. A more expensive way you can help fix the environment is to upgrade your appliances to energy-efficient ones or improve the insulation in your home by installing better windows, doors, and wall/attic insulation. This is expensive, but hopefully your government has some sort of grant to help cover the cost to make your home more environmentally friendly.

Slightly altering what you buy and how you shop will help as well. Just being conscious of where the products you are buying came from and choosing a more sustainable option if there is one will make a major difference. This includes everything like clothes, food, furniture, makeup, toiletries. There are products that cause a negative impact on the environment from their production due to mass deforestation, generate large amounts of GHGs and waste, and use or obtain raw resources in a non-sustainable way. There are innovative companies that have been trying to do their part to operate in a more responsible and sustainable way; unfortunately, we are not taking the time to figure out which brands are good and which are not. It can be tough to do the right thing, as it seems that the companies that do things properly have to charge more money for their product because it costs them more to do it properly; similarly, the cheaper products might be much more harmful to the environment, but are easier on the bank account. An issue that I run into quite frequently is choosing between buying a cheaper item, which I know if going to break in a month (like those gas station charger cables) or spending money on a better quality charging cable, which I know will last me much longer. Clearly, buying better quality stuff that lasts longer is better than just buying something that will only last you a couple of weeks before it ends up in the landfill; I am sure the GHGs and resources used to create large quantities of the cheap items are just devastating. Maybe the governments should enforce some sort of policy that if something is being sold, it needs to last with normal use for at least six months.

To reduce the amount of GHGs the manufacturers produce, we need to support better-made products, become more socially accepting of buying secondhand items, and get back to fixing things rather than just throwing them away. If we can get back to the good old days and relearned valuable life skills like fixing furniture and appliances, patching up and altering our clothes, making delicious meals out of food scraps and anything we could get our hands on, grow our own food etc., we would really make a huge dent out of the GHGs and waste production from the industrial and manufacturing industries as well as residential waste. If we just stop buying new things and find what we want through secondhand sources, we will reduce or eliminate the demand on manufacturing and industrial processes, and eliminate the GHGs from their processes. Both of those industries are just in business to provide us with things that we are putting a demand on; if we remove the demand, they will stop producing.

Another benefit of buying more secondhand things like clothing, furniture, appliances, etc., and keeping them in circulation is that it keeps the items out of the landfills. As the population increases, more land will not only be required for them to live on, but will be required to grow and produce food. Hopefully, the land will also be used for something like solar farms and waste-to-energy facilities. If we can figure out ways to repurpose and reuse the waste we generate for other people's uses, we would be setting ourselves up for success.

Companies have done an amazing job with marketing and making people feel like they need the newest, shiniest, and coolest stuff, but I think many people should take a step back and rethink what the difference between want and need is. According to NASA, we need food, water, air, and shelter, which means that everything else is just wants. Just take a second and figure out what you want more, a habitable planet which will provide you with food, water, and air, or do you want to continue to put a demand on mass production of things like electronics, clothes, makeup, and foreign foods? Do you NEED the newest phone? The answer is no. Does marketing and society make you feel like you need the newest phone? Probably. Companies pay top dollar to make you think like you need their newest products, but again, take a step back and think for yourself and ask yourself what do you want more, a healthy planet or to continue to put a demand on these companies mass producing items which you will probably be disposing of in a year when they try to sell you the same item, just in a different color or with a fifth camera and figure out how to take your money again. Just think of all the electronic waste you produce every time you buy a new smart watch, computer, phone, etc. There were materials mined, GHGs produced to manufacture your new device, which all had a negative impact on the environment, and now there is another device that contains harmful components like heavy metals and lithium-ion batteries to be disposed of. If you like funny movies, I recommend one called Idiocracy. It is from 2006, but it highlights some of these issues. Just being happy with what you have, it if breaks, try to fix it instead of replacing it, don't worry about social trends at getting the latest devices as it is more important for you to o your part to fix the environment then what other people think of you; this will be great for both the environment and your bank account. Doesn't that seem like a good situation?

If we make it cool to hold on to and continue to use your purchases for years (like your phone or computer), repair them instead of replace, and get to a point where people are showing off how they are still using the same phone from when they got it seven years ago instead of how they just got the newest version, we would all have a lot more money in our pockets and a healthier planet. If we shifted our demand from wanting new and different to being comfortable with using what we currently have, these major producers of GHGs in the manufacturing and industrial industries would be forced to change how they operate, and the environmental effects from their production would dramatically improve. Getting society to this point would be difficult. As mentioned before, companies pay a lot of money to try to convince you to buy their products, so to reverse that brainwashing, someone would have to pay for commercials, celebrity endorsements, sponsor sporting events and public functions to advertise the concept of not buying new things, keeping what they have and to buy used items to replace items which have actually reached the end of their lifecycle; I really doubt that this would happen because there is no financial gain from doing it, it would only help the planet.

The last bit of homework here for you is to try to cut back on your food waste. Looking at my food waste, it is mainly from buying a lot at one time because I don't want to go grocery shopping again anytime soon, and then forgetting about that bag of bok choy which has starting to leak orange goo in the back of the vegetable drawer, or the bag of potatoes in the pantry which have gone soft and starting to sprout little vines and I have to throw them away. I also create food waste because my eyes can be a lot bigger than my stomach, I put too much food on my plate, and I end up throwing the excess food away. Luckily, where I currently live, there is an organic food waste pickup every two weeks, just like garbage collection, so I can at least keep the food waste out of the landfill. How can I reduce my food waste? I can do this by planning my meals better and shopping specifically for what I will be making instead of just getting things that I think might be useful. I should do a better job of knowing what I have for food and try to plan around that as well; there are times when I make something and freeze it, then forget I have it until a year later when it is all frost-bitten and I must throw it away. I should also try to take smaller portions and go back for seconds if I am still hungry, instead of taking a big plate on the first round and not being able to finish it. I don't know what your situation is regarding food waste, but I used my current situation as an example to hopefully help you figure out solutions for your setup. What I was confused about is how it is okay for organic food to decompose in an organic landfill to make compost, but it's not okay for it to decompose at a regular landfill since it will produce the same amount of methane anyway, right? The reason why, which I just found out, is that the organic waste collection facilities are better equipped to capture the methane and deal with it. Many newer landfills should have a methane collection system, and I am hoping that the older landfills are getting methane collection systems installed, since the last thing we want is a pocket of methane to somehow combust in a landfill and create a massive fire fueled by burning plastics, rubber, and other garbage which is stockpiled in the landfill.

Choosing food which was locally produced will most likely be a more environmentally sustainable choice than buying food that are out of season and imported from somewhere that might not care about operating in a sustainable way. Reducing your meat consumption or switching your meat type to eggs, chickens, turkey, or even goats will greatly reduce the amount of GHGs produced for your food. Replacing the conventionally farmed meat in your diet to alternative proteins like lab made beef or plant based meat imitation foods would also be very beneficial I'm not talking about making major changes, just try to make it so that one or two dinners a week you eat a vegetarian meal, maybe a chili with lentils instead of ground beef, or something like that. You wouldn't even taste the difference if it's done decently. Little changes made by everyone could make a big difference. I know to you it might seem simple and insignificant and not worth doing, but if millions or more people all make these little changes, we would make a massive change for the better and might just pull through this. You just try to do your part and trust that others are doing theirs.

Grading

For those of you who like feedback, are competitive, or just like to know where you stand, I have put together a self-assessment/grading matrix which you can use to figure out where you stand regarding your environmental footprint and ways you can improve. This is a lighthearted exercise, don't take it too seriously; there are of course, always exceptions. If we can just get everyone operating at a "C" level, we should be able to pull through these environmental crises, but if you want to go the extra mile, help pick up some of the slack, I encourage you to go for the "A' grade. A+'s are for everyone who is honestly doing their very best, regardless of whether they check any of the boxes below. For the planet to get back to where it should be, we don't need to make absurd changes to our lives; many of us just need to slightly alter how we do things, and the great thing about many of these changes is that they will save you money, so it's a win-win. I'd say I am a solid C+, working my way up to a B with a long-term goal of an A.

Grade	What it takes to make the grades
A+	You are honestly doing your best to minimize your impact on the environment and fix the current environmental issues we are dealing with.
A	• Everything from B • Any energy you consume is from a "green" renewable resource. • Cook and heat with captured methane from your decomposing organics • Produce or sustainably hunt your own food, which is in season, and in a way that does not produce GHGs • Your home is completely built from environmentally friendly materials and in a way that naturally provides heating, cooling, and light. • All the possessions are second-hand / refurbished.

	• Traveling is kept to walking, biking, sailing, rowing (etc.)
B	• Everything from C+ • Lead by example, but do not force your ideas on others • Help with environmental restoration projects, either physical, fundraising, or donating • Promote changes within your community, like planting more trees to help reduce the temperature in cities and clean the air • Vote for political parties that are in favor of the Paris Accord or are going to create their own aggressive GHG reduction plan • Push your government to invest more in carbon sink restoration projects

C +	<ul><li>Everything from C</li><li>Learn new skills that help you get back to the old ways of fixing what is broken and making meals from scraps.</li><li>Fix your item instead of just replacing it, buy used items and restore them instead of buying new.</li></ul>
C	<ul><li>Try your best to minimize your consumption and demand for fossil fuels</li><li>Plan out your shopping to minimize food waste</li><li>Buy as much as possible second-hand instead of buying new just because you worry about what other people think of you.</li><li>Buy products that are produced sustainably and locally.</li><li>Make minor changes to your eating habits to reduce the demand for beef or livestock altogether.</li></ul>

C -	You want to do what you need to do your part, but haven't quite got around to doing it yet, but you want to and you will.
F	I feel the best way to get this point across is to direct you to a song written by Denis Leary, which depicts someone who would qualify for this grade; the song can be found here https://www.youtube.com/watch?v=UrgpZ0fUixs

Conclusion

Thinking through what we have gone through as well as what I notice daily, below is how I believe we should prioritize the steps of the planetary restoration if we get the resources to do it.

Reduce the amount of GHGs being generated.

- Reduce the number of wildfires
- early detection
- better technologies (fire suppressive projectiles)
- Better Forest Management
- Controlled burns
- Selective Logging

2. Invest in cleaner power-generating facilities
3. Use less power
4. Stop the thawing of the permafrost and the ice caps from melting
5. Reduce demand on the industrial and manufacturing industries
6. Change your eating habits, try less beef, more eggs, chicken, turkey, local goat or vegetables

1. Reducing the amount of incoming solar radiation and heat being absorbed
    - Blocking it before it enters our atmosphere
    - Reflect it back into the atmosphere and have some of it end up back in space, and at the same time prevent it from being absorbed by the ground, concrete, etc.
    - Green roofs, plant more trees.
    - Cool roads

2. Increase the amount of GHGs being removed

- Help phytoplankton populations increase by reducing microplastics and pollution
- Save forests from being cleared out by decreasing the demand for products from these areas and companies.
- Restore the wetlands, either do it yourself after enough research and obtaining free money from the government to do so, or raise money to help fund other restoration projects.
- Plant more trees

3. Reduce your waste production and demand for new products to be manufactured

- Reducing your food waste will make a big difference in both waste management and extreme climate change.
- Clean up the oceans
- Help developing countries set up proper waste management systems.
- Shop more at flea markets, garage sales, second-hand stores, and try not to buy new items; this will save you money and the environment. Setup a free item day for your neighborhood to help you get rid of something you don't need anymore and help provide your neighbor with something they might need, creating your own circular economy.
- Fix what is broken instead of replacing it.
- Worry more about your impact on the environment than social trends.
- Save your money and hold onto your possessions for as long as possible instead of just buying something new to replace them because it is new and shiny.

- Eat locally sourced, in-season foods. Let it be okay for your grocer to have empty shelves to help reduce food waste from food not purchased and going to waste

Well, that's it for me, for now, it's now up to you to do your part. The ideas are all here, ways you can tweak your life or completely revamp it, depending on what type of impact you are looking to have. I am sure that there are some of you who are tired of hearing about the environmental issues, and that is totally understandable. I am tired of hearing people just sitting around and talking about environmental issues but not doing anything about it. Let's just throw everything we have at it for a couple of years, let's take a time out on the race to Mars, and just focus on fixing the planet we are living on before we make it to and start inhabiting another planet, which we will most likely also ruin if we don't change our ways. It is true that if we figure out how to generate GHGs on Mars, it would help build an atmosphere there, keep the heat in, and make it more livable, but let's just focus on the immediate issue at hand here, please, and just get it over with. I want to live in a world where I don't hear about and witness major global issues from a degrading planet constantly, and to get to that point, we just all must put in some work. This book does not provide all of the answers since I had to balance the length of the book and when it gets published; I could spend more years writing this, but then it wouldn't be published for years and more time would have gone by, and the environmental issues would have probably gotten worse. If you are interested in doing your part to reduce your impact on the environment and help fix the environmental issues we are facing, go and conduct your own research, but make sure to only use reputable sources and not just go off of uneducated people's opinions.

It's risky putting all of our trust in the governments to fix these issues, I would love it if the Paris Agreement meets its goals, but I really don't have enough faith in many of the politicians to follow through with what they said they would do; I am used to them just saying what they need to say to get them another term in office and then pretending like they never said it. Also, I am sure it is tough running a country and especially a democratic one, you must figure out what most people want (and they all want different things) then you must balance the economy, employment, taxes, healthcare, national defense, international relations etc., and the environment might not be at or anywhere near the top of the list. The countries and politicians could even change their minds on the Paris agreement and pull out because they have to focus on something else, like a major multinational war, natural disaster, or pandemic. We need to rally the private sector; we have the plan, the only thing preventing the execution is the resources. A simple and easy solution would be for the major social media content creators and influencers to donate money and become environmental champions, then all of the content watchers just need to sit at home, watch the creators' / influencers' content to support them and raise more money to save the planet. Gen Z can literally help save the world by just sitting at home, watching TikTok and YouTube videos from the content creators and influencers who donate money to environmental projects, and then the content creators and influencers will go down in history as the ones who help save the world. Alternatively, the world might just need one or two champions, people who has accumulated large amounts of money and are looking to leave a legacy of being "the person who saved the world", or "the person who saved the world and the first person to get the spaceships to Mars"; doesn't that have a nice ring to it?

Just a quick update on myself, incase you were concerned about my employment status. I did end up filling a role with a top environmental firm in Canada as a Project Manager in their decommissioning/demolition / hazmat division for a few months, which is why the last part of this book was a little rushed. After a few months there, I was offered a Sr. Project Manager role to help improve the transmission portion of the clean energy power grid, which is where I currently am working while finishing up this book!

Sources

- Anthony, Leslie "Artic Permafrost is thawing. Here's what that means for Canada's North – and the world", Canadian Geographic, April 2, 2019, February 2024 <https://canadiangeographic.ca/articles/arctic-permafrost-is-thawing-heres-what-that-means-for-canadas-north-and-the-world/>
- Aquablu, 2024, *World's Biggest Plastic Polluters,* March 2024 <https://aquablu.com/stories/environment/worlds-biggest-plastic-polluters/>
- Ayala, Manuel Ochao, "Avacado: the "green gold" causing environment havoc" World Economic Forum, February 24, 2020, March 2024 <https://www.weforum.org/agenda/2020/02/avocado-environment-cost-food-mexico/#:~:text=Forest%20lands%20with%20diverse%20wildlife,it%20was%20lost%20to%20burning.>
- Baek, Seung H et al, "Impact of Climate on the Global Capacity for Enhanced Rock Weather on Croplands" *Earths Future, Volume 11, Issue 8,* August 14, 2023
- Bawa, A.S, Anilakumar, K.R., *Genetically Modified foods: safety, risks and public concerns – a review.* Springer, Journal of Food Science and Technology, December 19, 2012, <https://www.ncbi.nlm.nih.gov/pmc/articles/PMC3791249/>
- Bernet, Ross "How much CO2 does a tree absorb" One Tree Planted, July 25, 2023, February 2024 <https://onetreeplanted.org/blogs/stories/how-much-co2-does-tree-absorb>

- Bolles, Dana, “The Causes of Climate Change” National Aeronautics and Space Administration (NASA), March 2024, March 2024 <https://climate.nasa.gov/causes/>
- Brandon, Elissaveta, “The world has a concrete problem. These quirly alternatives want to be a solution.” Fast Company, October 18, 2023. March 2024 <https://www.fastcompany.com/90961431/the-world-has-a-concrete-problem-these-quirky-concrete-alternatives-want-to-be-a-solution>
- Carlowicz, Michael “Seeing Forests for the Trees and the Carbon: Mapping the World’s Forests in Three Dimensions” NASA Earth Observatory, January 9, 2012, February 2024 <https://Earthobservatory.nasa.gov/features/ForestCarbon#:~:text=Forests%20are%20considered%20one%20of,50%20percent%20of%20plant%20productivity.>
- Castle, S.E., Miller, D.C., Merten, N. *et al.* Evidence for the impacts of agroforestry on ecosystem services and human well-being in high-income countries: a systematic map. *Environ Evid* **11**, 10 (2022). https://doi.org/10.1186/s13750-022-00260-4
- Center for Climate and Energy Solutions, “Global Emissions”, *Climate Basics, Energy/Emissions Data,* March 2024 <https://www.c2es.org/content/international-emissions/>
- Centers for Disease Control and Prevention, U.S. Department of Health & Human Services, “Health Implications of Drought” January 16, 2020, February 2024 <https://www.cdc.gov/nceh/drought/implications.htm#:~:text=Drought%20can%20also%20cause%20long,mosquitoes%20breeding%20in%20stagnant%20water>

- Coast Adapt, March 2024 <https://coastadapt.com.au/ocean-acidification-and-its-effects>
- Courouble, Marianne et al, "Global Wetland Outlook – Special Edition 2021" Ramsar, 2021
- Clifford, Catherine "White House is pushing ahead research to cool Earth by Reflecting Back Sunlight" CNBC, Published October 13, 2022, Updated October 18, 2022, February 2024 <https://www.cnbc.com/2022/10/13/what-is-solar-geoengineering-Sunlight-reflection-risks-and-benefits.html>
- Dale, Pat, "Wetlands and human health: An Overview" Wetlands Ecology and Management 20(3), June 2012 <https://www.researchgate.net/publication/257673850_Wetlands_and_human_health_An_overview
- Ebert, Grace "A Japanese Forestry Technique Prunes a Tree Platform for More Sustainable Harvests" Colossal, October 27, 2020, April 2024 <https://www.thisiscolossal.com/2020/10/daisugi-trees-japan/>
- Environmental and Energy Study Institute, *Alternative Low / No Carbon Fuels,* February 2024 <https://www.eesi.org/topics/alternative-fuels/description>
- European Environmental Agency, *Efforts to transform Europe's economy into a circular one will require further bold action and strong implementation of existing measures.* February 2024 <https://www.eea.europa.eu/>
- European Space Agency, "Helium-3 mining on the lunar surface" April 2024 <https://www.esa.int/Enabling_Support/Preparing_for_t

he_Future/Space_for_Earth/Energy/Helium-3_mining_on_the_lunar_surface>

- FAO. 2022. Agricultural production statistics. 2000–2021. FAOSTAT Analytical Brief Series No. 60. Rome. <https://doi.org/10.4060/cc3751en>
- Filipenco, Daniil "World Waste: Statistics by country and short facts" Development Aid, March 7, 2023, February 2024, <https://www.developmentaid.org/news-stream/post/158158/world-waste-statistics-by-country>
- Fleming, Sean, "Circular economy examples – how IKEA, Burger King, Adidas and more are investing in a circular economy" World Economic Forum, December 8, 2020, March 2024 <https://www.weforum.org/agenda/2020/12/circular-economy-examples-ikea-burger-king-adidas/>
- Forest Stewardship Council, 2024, March 2024 <https://fsc.org/en>
- Forge, Frédéric, *Carbon Sequestration by Agricultural Soil,* Government of Canada Parliamentary Research Branch, January 30, 2001, <https://publications.gc.ca/Collection-R/LoPBdP/BP/prb0038-e.htm>
- Gagen, Mary et al, "Risking the Amazon: Why we need immediate action to reduce the tipping point risk" World Wildlife Federation, November 2022 <https://www.wwf.org.uk/our-reports/risking-amazon>
- GCFGlobal, "What is lab-grown meat?" April 2024 <https://edu.gcfglobal.org/en/thenow/what-is-labgrown-meat/1/>
- Geggel, Laura, "How often do ice ages happen" Live Science, March 25, 2017, February 2024, <https://www.livescience.com/58407-how-often-do-ice-ages-happen.html>

- Hannah Ritchie, Pablo Rosado and Max Roser (2020) - "Breakdown of carbon dioxide, methane and nitrous oxide emissions by sector" Published online at OurWorldInData.org. Retrieved from: 'https://ourworldindata.org/emissions-by-sector'
- Hannah Ritchie, Pablo Rosado and Max Roser (2020) - "Greenhouse gas emissions" Published online at OurWorldInData.org. Retrieved from: 'https://ourworldindata.org/greenhouse-gas-emissions'
- Heath Canada, Government of Canada, "Reducing Urban Heat Islands to Protect Health in Canada", April 29, 2020, March 2024 <https://www.canada.ca/en/services/health/publications/healthy-living/reducing-urban-heat-islands-protect-health-canada.html>
- Hirschchlag, Ally, "Plastic or Paper? The truth about drinking straws" BBC, November 5, 2023
- <https://www.bbc.com/future/article/20231103-plastic-or-paper-the-truth-about-drinking-straws> March 2024
- Igini, Martina "10 Shocking Statistics About Deforestation" Earth.org, January 21, 2023, March 2024 <https://Earth.org/statistics-deforestation/#:~:text=Since%201990%2C%20deforestation%20has%20robbed,in%20the%20world%20is%20spared.>
- Igini, Martina, "10 Concerning Fast Fashion Waste Statistics" Earth.org, August 21, 2023, March 2024 <https://Earth.org/statistics-about-fast-fashion-waste/>
- International Energy Agency, "An energy roadmap to carbon neutrality in China" February 2024 <https://www.iea.org/reports/an-energy-sector-roadmap-to-carbon-neutrality-in-china/executive-summary>

- International Energy Agency, "Net Zero Emissions by 2050 Scenario (NZE)" April 2024 <https://www.iea.org/reports/global-energy-and-climate-model/net-zero-emissions-by-2050-scenario-nze>
- Joint Economic Committee, "Climate-Exacerbated Wildfires cost the U.S. between $394 to $893 Billion each year in economic costs and damages." October 16, 2023, <https://www.jec.senate.gov/public/index.cfm/democrats/2023/10/climate-exacerbated-wildfires-cost-the-u-s-between-394-to-893-billion-each-year-in-economic-costs-and-damages#:~:text=The%20total%20cost%20of%20wildfires,to%20%24893%20billion%20each%20year.>
- Koluman, Nazan, "Goats and their role in Climate Change" Çukurova University, Agricultural Faculty, Department of Animal Science September 26, 2023, <https://www.sciencedirect.com/science/article/abs/pii/S0921448823001906
- Kim, Insoo et al "Decarbonizing the iron and steel industry: A systematic review of sociotechnical systems, technological innovations, and policy options" Elsevier, Energy Research & Social Science Volume 89, July 2022 <https://www.sciencedirect.com/science/article/abs/pii/S2214629622000706>
- Lea, Frederick M. et al, "Cement" Encyclopedia Britannica, March 10, 2024, <https://www.britannica.com/technology/cement-building-material>
- Lee, Hyun Jung et al. "Status of meat alternatives and their potential role in the future meat market – A Review." National Library of Medicine, National Center for Biotechnology Information, July 28, 2020,

<https://www.ncbi.nlm.nih.gov/pmc/articles/PMC7463075/>

- Limb, Lottie, "Forests are mostly to thank – with new tech only responsible for 0.1% of carbon dioxide removed so far, according to a new report." EuroNews.Green, January 19, 2023, March 2024 <https://www.euronews.com/green/2023/01/19/scientists-calculate-how-much-co2-is-being-removed-from-the-atmosphere-for-the-first-time>
- MacLeod, M.J., Hasan, M.R., Robb, D.H.F. *et al.* Quantifying greenhouse gas emissions from global aquaculture. *Sci Rep* **10**, 11679 (2020). <https://doi.org/10.1038/s41598-020-68231-8>
- Morrison, Rose, "Six Sustainable Alternatives to Traditional Concrete (and how to use them" Renewable Energy Magazine, June 5, 2023, <https://www.renewableenergymagazine.com/rose-morrison/six-sustainable-alternatives-to-traditional-concrete-and-20230605>
- Morsink, Kalila, "With Every Breath You Take, Thank the Ocean" Smithsonian, July 2017, February 2024, < https://ocean.si.edu/ocean-life/plankton/every-breath-you-take-thank-ocean>
- Moseman, Andrew; Rothman, Daniel, "How much carbon dioxide does the Earth Naturally Absorb?" Massachusetts Institute of Technology (MIT), January 26, 2024, March 2024 <https://climate.mit.edu/ask-mit/how-much-carbon-dioxide-does-Earth-naturally-absorb#:~:text=Altogether%20the%20planet%20absorbs%20and,billion%20tons%20of%20CO2.>
- National Aeronautics and Space Administration (NASA), "Climate Change", February 2024, <https://climate.nasa.gov/>

- National Geographic Society, "Formation of Earth" October 19, 2023, March 2024 <https://education.nationalgeographic.org/resource/formation-Earth/>
- National Oceanic and Atmospheric Administration, "Ocean Acidification" April 1, 2020, March 2024 <https://www.noaa.gov/education/resource-collections/ocean-coasts/ocean-acidification>
- National Oceanic and Atmospheric Administration, "Garbage Patches", March 21, 2024, March 2024, <https://marinedebris.noaa.gov/discover-marine-debris/garbage-patches>
- National Oceanic and Atmospheric Administration, "How does the ocean affect climate and weather on land?" March 2024 <https://oceanexplorer.noaa.gov/facts/climate.html#:~:text=The%20ocean%20influences%20weather%20and,globe%2C%20and%20driving%20weather%20systems.>
- Peters, Adele, "Could this new wind turbine design revolutionize renewable energy?" World Economic Forus, November 17, 2023, April 2023 <https://www.weforum.org/agenda/2023/11/new-wind-turbine-design-energy/#:~:text=In%20a%20field%20near%20the,the%20track%20and%20generate%20power.>
- Pragna, Pratap et al. "Climate Change and Goat Production: Eneric Methane Emission and Its Mitigation", National Library of Medicine, National Center for Biotechnology Information, December 7, 2018, <https://www.ncbi.nlm.nih.gov/pmc/articles/PMC6316019/#:~:text=Furthermore%2C%20goats%20emit%20less%20enteric,health%20and%20survival%20%5B6%5D.>

- Pütsep, Avely, “Which Trees Absorb the most carbon and why” Single Earth, October 11, 2021, April 2024 <https://www.single.Earth/blog/which-trees-absorb-the-most-carbon>
- Rainforest Alliance, “What is Sustainable Forestry” July 28, 2016, February 2024, <https://www.rainforest-alliance.org/insights/what-is-sustainable-forestry/>
- Rainfores Rescue, “Palm oil – deforestation for everyday products” April 2024 <https://www.rainforest-rescue.org/topics/palm-oil>
- Research Center for Engineering Ecology and Nonlinear Science, North China Electric Power University, Beijing 102206, China, *Phytoplankton Community Response to Environmental Factors along a Salinity Gradient in a Seagoing River, Tianjin, China* Microorganisms, December 27, 2022
- <https://www.ncbi.nlm.nih.gov/pmc/articles/PMC9864511/#:~:text=Numerous%20studies%20on%20phytoplankton%20communities,)%20and%20biological%20factors%20(e.g.%2C>
- Riebeek, Holli, “The Ocean’s Carbon Balance.” National Aeronautics and Space Administration (NASA) Earth Observatory, July 1, 2008, <https://Earthobservatory.nasa.gov/features/OceanCarbon#:~:text=(Graph%20by%20Robert%20Simmon.),water%20near%20the%20ocean's%20surface.>
- Samal, Kundan et al, “Pharmaceutical wastewater as Emerging Contaminants (EC): Treatment technologies, impact on environment and human health” Elsevier, Energy Nexus Volume 6, June 16, 2022, <https://www.sciencedirect.com/science/article/pii/S2772427122000390>

- Sensoneo, “Global Waste Index 2022” March 2024 <https://sensoneo.com/global-waste-index/>
- Shahbandeh, M, “Edible insects – statistics & facts” Statista, March 18, 2024, March 2024 <https://www.statista.com/topics/4806/edible-insects/#topicOverview
- Smithsonian, “Changes in Ultraviolet Radiation” Environmental Research Center, March 2024 <https://serc.si.edu/research/research-topics/global-change/changes-ultraviolet-radiation#:~:text=Ultraviolet%20radiation%20affects%20many%20processes,on%20nutrient%20cycling%20and%20decomposition.>
- Symington, Adam, “Mapping the World’s Forests: How Green is Our Globe” Visual Capitalist, January 20, 2030, February 2024 <https://www.visualcapitalist.com/cp/mapping-worlds-forests-how-green-is-our-globe/>
- Tenenbaum, Laura Faye, “Sun and sky, snow and ice. Reflections at the top of the world.” National Aeronautics and Space Administration (NASA), June 5, 2017, February 2024 <https://climate.nasa.gov/explore/ask-nasa-climate/2594/Sun-and-sky-snow-and-ice/#:~:text=Albedo%20is%20part%20of%20what,heat%20reflected%20back%20to%20space.>
- The Natural Resource Defense Council (NRDC), March 2024 <https://www.nrdc.org/>
- The Ocean Conservancy, February 2024 <https://oceanconservancy.org/trash-free-seas/plastics-in-the-ocean/>
- Turgeon, Andrew, Morse, Elizabeth, “Wind” National Geographic, October 19, 2023, March 2024

<https://education.nationalgeographic.org/resource/wind/>

- University Corporation for Atmospheric Research, "How Melting Artic Ice Affects Ocean Currents" 2024, April 2024 <https://scied.ucar.edu/learning-zone/climate-change-impacts/melting-arctic-sea-ice-and-ocean-currents#:~:text=The%20melting%20ice%20causes%20freshwater,caused%20the%20currents%20to%20slow.>
- U.S. Department of Agriculture, "Agroforestry" USDA Agroforestry Strategic Framework: Fiscal Year 2019-2024 <https://www.usda.gov/topics/forestry/agroforestry>
- U.S. Environmental Protection Agency (EPA), "Basic Ozone Layer Science" Ozone Layer Protection, October 7, 2021, February 2024 <https://www.epa.gov/ozone-layer-protection/basic-ozone-layer-science#:~:text=The%20ozone%20layer%20in%20the,that%20has%20several%20harmful%20effects.>
- U.S. Environmental Protection Agency (EPA), "Learn About Heat Islands", Heat Islands, August 28, 2023, March 2024 <https://www.epa.gov/heatislands/learn-about-heat-islands>
- U.S. Environmental Protection Agency (EPA), "Sources of Greenhouse Gas Emissions" Greenhouse Gas Emissions, April 11, 2024, February 2024 <https://www.epa.gov/GHGemissions/sources-greenhouse-gas-emissions#:~:text=Transportation%20(28%25%20of%202021%20greenhouse,ships%2C%20trains%2C%20and%20planes.>
- U.S. Environmental Protection Agency (EPA), "Climate Change Indicators: Sea Surface Temperature" Climate Change Indicators, November 1, 2023, March 2024

<https://www.epa.gov/climate-indicators/climate-change-indicators-sea-surface-temperature>

- U.S. Environmental Protection Agency (EPA), "Basic Information about Wetland Restoration and Protection" Wetlands, November 1, 2023, April 2024 <https://www.epa.gov/wetlands/basic-information-about-wetland-restoration-and-protection>
- U.S. Environmental Protection Agency (EPA), "Principles of Wetland Restoration" Wetlands, May 4, 2023, <https://www.epa.gov/wetlands/principles-wetland-restoration>
- Waste Management World, "Global Waste Index: Ranking the biggest waste polluters worldwide." November 8, 2023, March 2024 <https://waste-management-world.com/research/global-waste-index-2022-ranking-the-biggest-waste-polluters-worldwide/>
- Wear, Stephanie et al, *Sewage pollution, declining ecosystem health, and cross-sector collaboration,* Biological Conservation, Volume 255, March 2021 < https://www.sciencedirect.com/science/article/pii/S0006320721000628>
- Winters, Joseph, "All that plastic in the ocean is a climate change problem, too." Grist, October 17, 2022, <https://grist.org/science/all-that-plastic-in-the-ocean-is-a-climate-change-problem-too/>
- World Nuclear Association, "Nuclear Power in the World Today" March 2024, April 2024 https://world-nuclear.org/information-library/current-and-future-generation/nuclear-power-in-the-world-today.aspx
- World Wildlife Federation, "Fight Climate Change by Preventing Food Waste" April 2024 <https://www.worldwildlife.org/stories/fight-climate-change-by-preventing-food-

waste#:~:text=But%20wasted%20food%20isn't,more%20potent%20than%20carbon%20dioxide.>

- YaleEnvironment360, "How adding Rock Dust to Soil could help get Carbon into the ground" E360 Digest, September 2, 2021, March 2024 <https://e360.yale.edu/features/how-adding-rock-dust-to-soil-can-help-get-carbon-into-the-ground>
- Yale Environment360, "Spreading Rock Dust on Farmland Has Potential to Draw Down Huge Sums of Carbon Dioxide" E360 Digest, August 15, 2023, February 2024 <https://e360.yale.edu/digest/enhanced-rock-weathering-climate-change>

www.ingramcontent.com/pod-product-compliance
Lightning Source LLC
LaVergne TN
LVHW052113210926
848482LV00025B/582
9798891226227